MODELING AND VISUALIZATION OF COMPLEX SYSTEMS AND ENTERPRISES

STEVENS INSTITUTE SERIES ON COMPLEX SYSTEMS AND ENTERPRISES

William B. Rouse, Series Editor

WILLIAM B. ROUSE
Modeling and Visualization of Complex Systems and Enterprises

MODELING AND VISUALIZATION OF COMPLEX SYSTEMS AND ENTERPRISES

Explorations of Physical, Human, Economic, and Social Phenomena

WILLIAM B. ROUSE

Published by John Wiley & Sons, Inc., Hoboken, New Jersey
Published simultaneously in Canada

For general information on our other products and services or for technical support, please contact our Customer Care Department within the United States at (800) 762-2974, outside the United States at (317) 572-3993 or fax (317) 572-4002.

Wiley also publishes its books in a variety of electronic formats. Some content that appears in print may not be available in electronic formats. For more information about Wiley products, visit our web site at www.wiley.com.

Library of Congress Cataloging-in-Publication Data:

Rouse, William B.
  Modeling and visualization of complex systems and enterprises : explorations of physical, human, economic, and social phenomena / William B. Rouse.
        pages cm. – (Stevens Institute Series on Complex Systems and Enterprises)
    Includes bibliographical references and index.
    ISBN 978-1-118-95413-3 (cloth)
  1.   System theory–Mathematical models. 2.   System analysis. 3.   System theory–Philosophy.
I. Title.
    Q295.R68 2915
    003–dc23

                                                                                2015007877

Printed in the United States of America

10 9 8 7 6 5 4 3 2 1

1   2015

CONTENTS

2 Overall Methodology   27

3 Perspectives on Phenomena   43

9   Computational Methods and Tools   **209**

PREFACE

The seeds for this book were sown almost 50 years ago when, as a budding engineer at Raytheon, I was given the assignment of determining the optimal number and types of spare parts to take on a submarine for the sonar system. I had to integrate reliability, maintainability, and availability models into an overall mission simulation to assess the effectiveness of alternative spare parts plans.

I have been immersed in mathematical and computational modeling ever since. For several years, I focused on operation and maintenance of complex vehicle and process systems, with particular emphasis on model-based decision aiding and training of personnel in these systems.

I next addressed design of new products and strategic business management. We developed a suite of software tools to support these processes and worked with over 100 companies and thousands of executives and senior managers. These intense experiences led to fascination with enterprises and the great difficulties they had with recognizing needs to change and especially accomplishing change.

Over much of the past decade, I have been immersed in transformation of the healthcare industry, often through collaborations with the National Academy of Engineering and the Institute of Medicine. It quickly became apparent that the poor performance of the US system could only be coherently understood by looking at the interactions of the multiple levels of the system.

Multi-level models of enterprise systems soon became a paradigm I applied to a range of types of enterprises. The need to better understand these types of models and better inform modeling methodology motivated this book. The material presented in this book is intended to help the many stakeholders in such modeling endeavors to understand the intricacies of this approach and achieve greater success.

I am grateful to many people who strongly influenced my thinking in planning and preparing this book. Mike Pennock has been an amazing sounding board and source of critiques and ideas. He and I are currently pursuing several of the research issues outlined later in this book. Conversations with John Casti, as well as several of his books, also have been very helpful.

Ken Boff, Alec Garnham, Bill Kessler, and Hal Sorenson helped me to understand the aerospace and defense industry. Bill Beckenbaugh, Jim Prendergast, and Dennis Roberson were my guides in the electronics and semiconductor industry. Denis Cortese, Mike Johns, and Bill Stead provided insights into healthcare delivery. Brainstorming with Alan Blumberg and Alex Washburn, as well as Michael Bruno and Dinesh Verma, regarding urban resilience broadened my perspective substantially.

The research of Rob Cloutier, Babak Heydari, Jose Ramirez-Marquez, and Steve Yang at Stevens provided interesting insights. Rahul Basole, Doug Bodner, Leon McGinnis, and Nicoleta Serban at Georgia Tech were kindred spirits in modeling pursuits. Research sponsors Kristin Baldwin, Judith Dahmann, and Scott Lucero often asked insightful questions and offered suggestions that shaped my thinking.

WILLIAM B. ROUSE
Hoboken, NJ
August 2014

1

INTRODUCTION AND OVERVIEW

Addressing complex systems such as health-care delivery, sustainable energy, financial systems, urban infrastructures, and national security requires knowledge and skills from many disciplines, including systems science and engineering, behavioral and social science, policy and political science, economics and finance, and so on. These disciplines have a wide variety of views of the essential phenomena underlying such complex systems. Great difficulties are frequently encountered when interdisciplinary teams attempt to bridge and integrate these often-disparate views.

This book is intended to be a valuable guide to all the disciplines involved in such endeavors. The central construct in this guide is the notion of phenomena, particularly the essential phenomena that different disciplines address in complex systems. Phenomena are observed or observable events or chains of events. Examples include the weather, climate change, traffic congestion, aggressive behaviors, and cultural compliance. A team asked to propose policies to address the problem of overly aggressive motorist behaviors during inclement weather in the evening rush hour might have to consider the full range of these phenomena.

Traditionally, such problems would be decomposed into their constituent phenomena, appropriate disciplines would each be assigned one piece of the

Modeling and Visualization of Complex Systems and Enterprises:
Explorations of Physical, Human, Economic, and Social Phenomena, First Edition. William B. Rouse.
© 2015 John Wiley & Sons, Inc. Published 2015 by John Wiley & Sons, Inc.

puzzle, and each disciplinary team would return from their deliberations with insights into their assigned phenomena and possibly elements of solutions. This reductionist approach often leads to inferior solutions compared to what might be achieved with a more holistic approach that explicitly addresses the interactions among phenomena and central trade-offs underlying truly creative solutions. This book is intended to enable such holistic problem solving.

Five themes are woven throughout this book.

- Understanding the essential phenomena underlying the overall behaviors of complex systems and enterprises can enable improving these systems.
- These phenomena range from physical, behavioral, and organizational, to economic and social, all of which involve significant human components.
- Specific phenomena of interest and how they are represented depend on the questions of interest and the relevant domains or contexts.
- Visualization of phenomena and relationships among phenomena can provide the basis for understanding where deeper exploration is warranted.
- Mathematical and computational models, defined *very* broadly across disciplines, can enable the necessary deeper understanding.

This chapter proceeds as follows. We first consider the nature of a range of perspectives on systems. This begins with an exploration of historical perspectives, drawing upon several disciplines. We then consider the nature of complexity and complex systems. This leads to elaboration of the contrast between complex and complicated systems and the notion of complex adaptive systems. We then consider systems practice over the past century. This background is intended to provide a well-informed foundation that will enable digesting the material discussed in later chapters.

SYSTEMS PERSPECTIVES

It is useful to reflect on the roots of systems thinking. This section begins with a discussion of the systems movement. We then elaborate the philosophical underpinnings of systems thinking. Finally, we review a range of seminal concepts. Brief sketches of these concepts are presented here; they are elaborated in greater depth in later chapters.

Systems Movement

The systems movement emerged from the formalization of systems theory as an area of study during and following World War II, although it can be argued that the physicists and chemists of the 19th century contributed to the foundations of this area. Before delving into the ideas emerging in the 1940s and beyond, it is important to distinguish four aspects of the systems movement:

- *Systems Thinking* is the process of understanding how things influence one another within a whole and represents an approach to problem solving that views "problems" as components of an overall system.
- *Systems Philosophy* is the study of systems, with an emphasis on causality and design. The most fundamental property of any system is the arbitrary boundary that humans create to suit their own purposes.
- *Systems Science* is an interdisciplinary field that studies the characteristics of complex systems in nature and society, to develop interdisciplinary foundations, which are applicable in a variety of areas, such as engineering, biology, medicine, and economics.
- *Systems Engineering* is an interdisciplinary field focused on identifying how complex engineering undertakings should be designed, developed, and managed over their life cycles.

Contrasting these four aspects of systems, it is important to recognize that different disciplines tend to see "systems" quite differently, for the most part due to the varying contexts of interest (Adams et al., 2014). Thus, a systems scientist studying marsh ecosystems and a systems engineer designing and developing the next fighter aircraft will, from a practical perspective at least, have much less in common than the term "system" might lead one to expect. The key point is that systems exist in contexts and different contexts may (and do) involve quite disparate phenomena.

Philosophical Background

There are many interpretations of what system thinking means and the nature of systems thinkers. Some are inclined toward model-based deduction, while others are oriented toward data-driven inference. The former extol the deductive powers of Newton and Einstein, while the latter are enamored with the inferential capabilities of Darwin. These different perspectives reflect different epistemologies.

The study of epistemology involves the questions of what is knowledge, how can it be acquired, and what can be known. The empiricism branch of epistemology emphasizes the value of experience. The idealism branch sees knowledge as innate. The rationalism branch relies on reason. The constructivism branch seeks knowledge in terms of creation. These branches differ in terms of how they represent knowledge, in particular how this knowledge is best modeled and simulated (Tolk, 2013).

There are many possible ways of thinking about complex systems and enterprises (Rouse, 2005, 2007). Systems paradigms for representation of knowledge include hierarchical mappings, state equations, nonlinear mechanisms, and autonomous agents (Rouse, 2003). For hierarchical mappings, complexity is typically due to large numbers of interacting elements. With uncertain state equations, complexity is due to large numbers of interacting state variables and significant levels of uncertainty. Discontinuous, nonlinear mechanisms attribute complexity to departures from the expectations stemming from continuous, linear phenomena. Finally, autonomous agents generate complexity via the reactions of agents to each other's behavior and lead to emergent phenomena. The most appropriate choice among these representations depends on how the boundaries of the system of interest are defined (Robinson et al., 2011).

Horst Rittel argued that the choice of representation is particularly difficult for "wicked problems" (Rittel & Webber, 1973). There is no definitive formulation of a wicked problem. Wicked problems have no stopping rule – there is always a better solution, for example, "fair" taxation and "just" legal systems. Solutions to wicked problems are not true or false, but good or bad. There is no immediate or ultimate test of a solution to a wicked problem. Wicked problems are not amenable to trial-and-error solutions. There is no innumerable (or an exhaustively describable) set of potential solutions and permissible operations. Every wicked problem is essentially unique. Every wicked problem can be considered a symptom of another problem. Discrepancies in representations can be explained in numerous ways – the choice of explanation determines the nature of problem's resolution. Problem solvers are liable for the consequences of the actions their solutions generate. Many real-world problems have the aforementioned characteristics.

The notion of wicked problems raises the possibility of system paradoxes (Baldwin et al., 2010). Classic paradoxes include whether light is a particle or a wave. Contemporary paradoxes include both collaborating and competing with the same organization. The conjunction paradox relates to the system including element A and element not A. The biconditional paradox holds if A implies B and B implies A. For the equivalence paradox, system elements have contradictory qualities. With the implication paradox, one or more system elements lead to its own contradiction. The disjunction para-dox involves systems that are more than the sum of their parts. Finally, the perceptual paradox reflects perceptions of a system that are other than reality.

Finally, there are fundamental theoretical limits as to what we can know about a system and its properties (Rouse and Morris, 1986; Rouse et al., 1989; Rouse and Hammer, 1991). There are limits of system information process-ing capabilities (Chaitin, 1974), limits to identifying signal processing and symbol processing models, limits of validating knowledge bases underlying intelligent systems, and limits of accessibility of mental models in terms of forms and content of representations. The implication is that models are inher-ently approximations of reality and may be biased and limited in significant ways. This topic is pursued in more depth in Chapter 5.

This broad – and very brief – review of the philosophical underpinnings of the systems arena leads to two very important observations. First, the range of disciplines involved and the variety of formalisms they employ has led to a lack of crispness in the nature of the field. Second, this state of affairs can, to a great extent, be attributed to the very wide range of phenomena of interest, for example, biological cells to urban infrastructures to macroeconomic policies. Chapters 4–7 address this variety by partitioning it into classes of phenomena, then recombining these elements in Chapters 8–10.

Seminal Concepts – Systems Science

The experiences of the problem-driven research in World War II led many now-notable researchers to develop new concepts, principles, models, methods, and tools for specific military problems that they later generalized to broader classes of phenomena. The systems theorists included Norbert Wiener (1961), who generalized control theory into the concept of cybernet-ics. Weiner defined cybernetics as the study of control and communication in the animal and the machine. Studies in this area focus on understanding and defining the functions and processes of systems that have goals and that participate in circular, causal chains that move from action to sensing to comparison with desired goals and back again to action. Concepts studied include, but are not limited to, learning, cognition, adaptation, emergence, communication, efficiency, and effectiveness. Later extensions of control theory include optimal state filtering (Kalman, 1960) and optimal control (Bellman, 1957; Pontryagin et al, 1962).

Shannon (1948) developed information theory to address the engineering problem of the transmission of information over a noisy channel. The most important result of this theory is Shannon's coding theorem, which estab-lishes that, on average, the number of bits needed to represent the result of an uncertain event is given by its entropy, where entropy is a measure of the uncertainty associated with a random variable. In the context of information theory, the term refers to Shannon entropy, which quantifies the expected value of the information contained in a message, typically measured in binary digits or bits. Shannon's noisy-channel coding theorem states that reliable communication is possible over noisy channels provided that the rate of communication is below a certain threshold, called the channel capacity. The channel capacity can be approached in practice by using appropriate encoding and decoding systems.

Ashby (1952, 1956) added the Law of Requisite Variety to the canon. Put succinctly, only variety can destroy variety. More specifically, if a system is to be fully regulated, the number of states of its control mechanism must be greater than or equal to the number of states in the system being controlled. Thus, in order for an enterprise to reduce the variety manifested by its environment to yield less varied products and services, it must have sufficient variety in its business processes.

Bertalanffy (1968) developed General Systems Theory over several decades, with particular interest in biological and open systems, that is, those that continuously interact with their environments. The areas of systems science that he included in his overall framework encompass cybernetics; theory of automata; control theory; information theory; set, graph, and network theory; decision and game theory; modeling and simulation; and dynamical systems theory – in other words, virtually all of systems science. Bertalanffy includes consideration of systems technology including control technology, automation, computerization, and communications. Had the field of artificial intelligence existed in his time, that area would have surely been included as well. As is often the case with grand generalizations, it is often difficult to argue with the broad assertions but sometimes not easy to see the leverage gained.

Ackoff (1971) coined the term "system of systems" that has gained great currency of late. He recognized that organizations could be seen as systems. In this context, he outlined a classification of systems (self-maintaining, goal-seeking, multigoal seeking, purposive system), and elaborated the notions of system state, system changes, and system outcomes, where outcomes are seen as the consequences of system responses, not just the response variables in themselves. He further elaborated organizational systems as being variety increasing or variety decreasing, and also discusses adaptation and learning.

Seminal Concepts – Economics/Cognition

It may seem odd to group economics with cognition. However, much seminal thinking arose from people who studied behavioral and social phenomena associated with economic processes. Nobel Prize winner Kenneth Arrow (Arrow, 1951; Arrow and Debreu, 1954) developed social choice theory, the associated impossibility theorem, equilibrium theory, and the economics of information. Nobel Prize winner Herbert Simon (1957, 1962) studied bounded rationality, satisficing versus optimizing, behavioral complexity as a reflection of environmental complexity, human information processing, and artificial intelligence. Nobel Prize winner Daniel Kahneman (2011), with his colleague Amos Tversky, studied human decision-making biases and heuristics for several decades. Finally, George Miller (1956) contributed to cognitive psychology, cognitive science, psycholinguistics (which links language and cognition), and studies of short-term memory – coming up with oft-cited "magic number seven."

This body of work provides important insights into complex systems laced with behavioral and social phenomena (as well as into how to win a Nobel Prize in Economics). Put simply, the classical notional of "economic man" as a completely rational, decision maker who can be counted on to make optimal choices is often a wildly idealistic assumption. The phenomena studied by Arrow, Simon, Kahneman, and Miller make classical mathematical economics quite difficult. On the other hand, these phenomena can make agent-based simulations quite important. In Chapters 5, 7, and 9, human decision making and problem solving are considered in some depth, with many concepts traceable back to the seminal thinkers discussed in this section.

Seminal Concepts – Operations Research

Operations research (OR) emerged from World War II and efforts to study and improve military operations. Philip Morse was a pioneer in the research philosophy of immersing problem solvers in the complex domains where solutions are sought. The key element was the emphasis on research in operational contexts rather than just study of mathematical formalisms. Morse and Kimball (1951) and Morse (1958) authored the first books in the United States in this area, and went on to publish an award-winning book on the application of OR to libraries (Morse, 1968).

C. West Churchman was internationally known for his pioneering work in OR, system analysis, and ethics. He was recognized for his then radical concept of incorporating ethical values into operating systems (Churchman, 1971). Ackoff received his doctorate in philosophy of science in 1947 as Churchman's first doctoral student (Ackoff et al., 1957). He became one of the most important critics of the so-called "technique-dominated Operations Research" and proposed more participative approaches. He argued that any human-created system can be characterized as a "purposeful system" when its "members are also purposeful individuals who intentionally and collectively formulate objectives and are parts of larger purposeful systems" (Ackoff & Emery, 1972).

More recently, OR has come to be dominated by applied mathematicians who pursue mathematical techniques as ends in themselves. The quest for provably optimal solutions of problems has resulted in problems being scaled down, often dramatically, to enable analytical proofs of optimality. The constructs of theorems and proofs have often displaced the intention to actually solve realistically complex problems. The value of immersing researchers in complex operational domains has often come to be discounted as impractical by the researchers themselves.

Seminal Concepts – Sociology

Talcott Parsons was one of the first social scientists to become interested in systems approaches. He developed action theory, the principle of voluntarism, understanding of the motivation of social behavior, the nature of social evolution, and the concept of open systems (Parsons, 1937, 1951a, 1951b; Parsons and Smelser, 1956). This very much set the stage for the emergence of socio-technical systems as an area of study in its own right.

The idea of work systems and the socio-technical systems approach to work design was originated by Trist, Emery, and colleagues (Trist & Bamforth, 1951; Emery & Trist, 1965, 1973). This included research on participative work design structures and self-managing teams. It also led to a deep appreciation of the roles of behavioral and social phenomena in organizational outcomes and performance.

COMPLEXITY AND COMPLEX SYSTEMS

The six archetypal problems that are introduced in Chapter 2 can be viewed as complex systems problems. This begs the question of the meaning of complexity and complex systems. There is a range of differing perspectives on the nature of complex systems (Rouse, 2003, 2007; Rouse & Serban, 2011). In particular, different disciplines, in part due to the contexts in which they work, can have significantly varying views of complexity and complex systems (Rouse et al., 2012).

Several concepts are quite basic to understanding complex systems. One key concept is the dynamic response of a system as a function of structural and parametric properties of the system. The nature of the response of a system, as well as the stability and controllability of this response, is a central concern. Many OR studies focus on steady-state behavior, while economics research addresses equilibrium behavior. However, transient behaviors – whether of the weather or the financial system – are often the most interesting and sometimes the most damaging.

Another basic concept is uncertainty about a system's state. The state of a system is the quantities/properties of the system whose knowledge, along with future inputs, enables prediction of future values of this set of variables. Uncertainty of system state limits the effectiveness of control strategies in assuring system performance. State estimation – filtering, smoothing, and prediction – is an important mechanism for obtaining the best information for controlling a complex system. Related topics include the value of information and performance risks, for example, consequences of poor performance.

It is useful to differentiate the notions of "system" and "complex system" (Rouse, 2003). A system is a group or combination of interrelated, interdependent, or interacting elements that form a collective entity. Elements may include physical, behavioral, or symbolic entities. Elements may interact physically, computationally, and/or by exchange of information. Systems tend to have goals/purposes, although in some cases, the observer ascribes such purposes to the system from the outside so to speak.

Note that a control system could be argued to have elements that interact computationally in terms of feedback control laws, although, one might also argue that the interaction takes place in terms of the information that embodies the control laws. One could also describe the control function in terms of physical entities such as voltages and displacements. Thus, there are (at least) three different representations of the same functionality – hence, the "and/or" in the definition.

A complex system is one whose *perceived* complicated behaviors can be attributed to one or more of the following characteristics: large numbers of elements, large numbers of relationships among elements, nonlinear and discontinuous relationships, and uncertain characteristics of elements and relationships. From a functional perspective, the presence of complicated behaviors, independent of underlying structural features, may be sufficient to judge a system to be complex. Complexity is perceived because apparent complexity can decrease with learning.

More specifically, system complexity tends to increase with

- Number of elements
- Number of relationships
- Nature of relationships
  - Logical: AND versus OR and NAND
  - Functional: linear versus nonlinear
  - Spatial: lumped versus distributed

- Structural: for example, feedforward versus feedback
- Response: static versus dynamic
- Time constant: (not too) fast versus (very) slow
- Uncertainty: known properties versus unknown properties
- Knowledge, experience, and skills
  - Relative to all of the above
  - Relative to observer's intentions

Of course, the preceding list begs the question of whether the elements of a system are knowable. For example, this list is of limited use in describing a city, except perhaps for the utility infrastructures. Thus, as elaborated later, we have to differentiate complex and complicated systems.

The issue of intentions is summarized in Figure 1.1 (Rouse, 2007). If one's intention is simply to classify as observed object as an airplane, the object is not particularly complex. If one wanted to explain why it is an airplane, the complexity of an explanation would certainly be greater than that of a classification. For these two intentions, one is simply describing an observed object.

If one's intention is to predict the future state of the airplane, complexity increases substantially as one would have to understand the dynamic nature of the object, at least at a functional level but perhaps also at a structural level. Control requires a higher level of knowledge and skill concerning input–output relationships. Intentions related to detection and diagnosis require an even greater level of knowledge and skill concerning normal

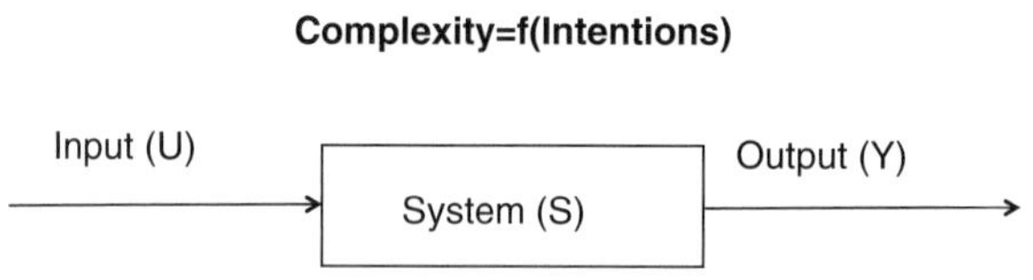

Complexity=f(Intentions)

Intention	Example
Classification	"It's an instance of type S".
Explanation	"It's type S because"....
Prediction	"It's future output will be Y".
Control	"If input is U, it's output will be Y".
Detection	"It's output is not Y, but should be".
Diagnosis	"It's output is not Y because"...

Figure 1.1   Relationship of Complexity and Intentions

and off-normal behaviors in terms of symptoms, patterns, and structural characteristics of system relationships. The overall conclusion is that the complexity of a system cannot be addressed without considering the intentions associated with addressing the system.

This observation is fundamental, and often hotly debated. It argues that complexity involves a relationship between an observer and an object or system. The implication is that there is no absolute complexity independent of the observer. In other words, the complexity of a system depends on why you asked the question, as well as your knowledge and skill relative to the system of interest.

COMPLEX VERSUS COMPLICATED SYSTEMS

Snowden and Boone (2007) have argued that there are important distinctions that go beyond those outlined earlier. Their Cynefin Framework includes simple, complicated, complex, and chaotic systems. Simple systems can be addressed with best practices. Complicated systems are the realm of experts. Complex systems represent the domain of emergence. Finally, chaotic systems require rapid responses to stabilize potential negative consequences. The key distinction with regard to the types of contexts discussed in this book is complex versus complicated systems. There is a tendency, they contend, for experts in complicated systems to perceive that their expertise, methods, and tools are much more applicable to complex systems than is generally warranted.

Poli (2013) also elaborates the distinctions between complicated and complex systems. Complicated systems can be structurally decomposed. Relationships such as listed earlier can be identified, either by decomposition or, in some cases, via blueprints. "Complicated systems can be, at least in principle, fully understood and modeled." Complex systems, in contrast, cannot be completely understood or definitively modeled. He argues that biology and all the human and social sciences address complex systems.

Poli also notes that problems in complicated systems can, in principle, be solved. The blueprints, or equivalent, allow one to troubleshoot problems in complicated systems. In contrast, problems in complex systems cannot be solved in the same way. Instead, problems can be influenced so that unacceptable situations are at least partially ameliorated.

Alderson and Doyle (2010) also discuss contrasting views of complexity and distinguish the constructs of simplicity, disorganized complexity, and organized complexity, drawing upon several of the post World War II seminal thinkers discussed earlier. With simplicity, "Questions of interest can be posed using models that are readily manageable and easy to describe, theorem statements are short, experiments are elegant and easy to describe, and require minimal interpretation. Theorems have simple counterexamples or short proofs, algorithms scale, and simulations and experiments are reproducible with predictable results."

Disorganized complexity "Focuses on problems with asymptotically infinite dimensions and develops powerful techniques of probability theory and statistical mechanics to deal with problems. As the size of the problem and the number of entities become very large, specific problems involving ensemble average properties become easier and more robust and statistical methods apply."

They illustrate this view with a discussion of the new science of complex networks, "which emphasizes emergent fragilities in disorganized systems. Proponents of this paradigm view architecture as graph topography." Their view of the Internet, for example, is "random router and web graphs without system-level functions other than graph connectivity, architecture solely in terms of graph topology, and components as homogeneous functionless links and nodes."

In contrast, organized complexity "Addresses problems where organization is an essential feature, which include biological systems, urban systems, and technological systems. Organized complexity manages the fragility–complexity spiral." For example, it views architecture as involving layering and protocols, rather than just the random connections of disorganized complexity.

The distinctions articulated by these authors are well taken. Complicated systems have often been designed or engineered. There are plans and blueprints. There may be many humans in these systems, but they are typically playing prescribed roles. In contrast, complex systems, as they define them, typically emerge from years of practice and precedent. There are no plans and blueprints. Indeed, much research is often focused on figuring out how such systems work. A good example is human biology.

The nature of human and social phenomena within such systems is a central consideration. Systems where such phenomena play substantial roles are often considered to belong to a class of systems termed complex adaptive systems (Rouse, 2000, 2008). Systems of this type have the following characteristics:

- They tend to be *nonlinear, dynamic* and do not inherently reach fixed equilibrium points. The resulting system behaviors may appear to be random or chaotic.

- They are composed of *independent agents* whose behaviors can be described as based on physical, psychological, or social rules, rather than being completely dictated by the physical dynamics of the system.
- Agents' needs or desires, reflected in their rules, are not homogeneous and, therefore, their *goals and behaviors are likely to differ or even conflict* – these conflicts or competitions tend to lead agents to adapt to each other's behaviors.
- Agents are *intelligent and learn* as they experiment and gain experience, perhaps via "meta" rules, and consequently change behaviors. Thus, overall system properties inherently change over time.
- Adaptation and learning tends to result in *self-organization* and patterns of behavior that emerge rather than being designed into the system. The nature of such emergent behaviors may range from valuable innovations to unfortunate accidents.
- There is *no single point(s) of control* – system behaviors are often unpredictable and uncontrollable, and no one is "in charge." Consequently, the behaviors of complex adaptive systems usually can be influenced more than they can be controlled.

As might be expected, understanding and influencing systems having these characteristics creates significant complications. For example, the simulation of such models often does not yield the same results each time. Random variation may lead to varying "tipping points" among stakeholders for different simulation runs. These models can be useful in the exploration of leading indicators of the different tipping points and in assessing potential mitigations for undesirable outcomes. This topic is addressed in more detail later.

SYSTEMS PRACTICE

The evolving collection of approaches to understanding and influencing complex systems and enterprises can be termed systems practice. The development of systems practice has a rich history.

- During the 1900–1920s, Henry Gantt (1861–1919), Frederick Taylor (1856–1919), and Frank Gilbreth (1868–1924) pioneered scientific management.
- Quality assurance and quality control emerged in the 1920–1930s, led by Walter Shewhart (1891–1967).
- Peter Drucker (1909–2005) and Chester Barnard (1886–1961) formalized corporate operations management in the 1940–1950s.

- During and following World War II, Philip Morse (1903–1985), C. West Churchman (1913–2004), George Dantzig (1914–2005), and Russell Ackoff (1919–2009) were leading thinkers in OR.
- Stafford Beer (1926–2002) articulated the foundations of management cybernetics in the 1960–1970s.
- W. Edwards Deming (1900–1993) and Joseph Juran (1904–2008) brought total quality management to the United States in the 1970–1980s.
- Michael Hammer (1948–2008) and James Champy led the wave of business process reengineering in the 1990s.
- Taiichi Ohno's (1912–1990) innovations in six sigma and lean production gained traction in the United States in the 1990–2000s.
- Most recently, Daniel Kahneman has led the way for behavioral economics in the 2010s.

The cornerstone of systems practice is usually considered to be systems thinking, which has been characterized in a variety of ways, depending on the analytic paradigms of interest, for example, (Checkland, 1993; Weinberg, 2001; Jackson, 2003; Meadows, 2008; Gharajedaghi, 2011). Over more than a century, systems thinking tried to become increasingly rigorous, focusing on mathematics, statistics, and computation. During the 1960–1970s, many thought leaders began to recognize that forcing all phenomena into this mold tended to result in many central phenomena being assumed away to allow for the much-sought theorems and proofs to be obtained. In particular, behavioral and social phenomena associated with complex systems were simplified by viewing humans as constrained but rational decision makers who always made choices that optimized the objective performance criteria – which were linear, if lucky.

The reaction, particularly in the United Kingdom, to such obviously tenuous assumptions was the emergence of the notion of hard versus soft systems thinking (Pidd, 2004). Table 1.1 contrasts these two points of view. Hard systems thinking seeks quantitative solutions of mathematical models that are assumed to be valid representations of the real world and, consequently, will inherently be embraced once they are calculated. Soft systems thinking sees modeling as a means for exploration and learning via intellectual and inherently approximate constructs open to discussion and debate.

Table 1.2 contrasts systems approaches (Jackson, 2003). Hard systems thinking represents but one cell in this table. Other methods are much less "closed form" in orientation, relying more on simulation as well as participative mechanisms. The keys for these latter mechanisms are insights and consensus building.

TABLE 1.1   Hard versus Soft Systems Thinking (Pidd, 2004)

Hard Systems Thinking	Soft Systems Thinking
Oriented to goal seeking	Oriented to learning
Assumes the world contains systems that can be "engineered"	Assumes the world is problematical but can be explored using models or purposeful activity
Assumes systems models to be models of the world	Assumes systems models to be intellectual constructs to help debate
Talks the language of problems and solutions	Talks the language of issues and accommodations
Philosophically positivistic	Philosophically phenomenological
Sociologically functionalist	Sociologically interpretative
Systematicity lies in the world	Systematicity lies in the process of inquiry into the world

TABLE 1.2   Systems Approaches (Jackson, 2003)

		Participants		
		Unitary	Pluralist	Coercive
Systems	Simple	Hard systems thinking	Soft systems approaches	Emancipatory systems thinking
	Complex	System dynamics Organizational cybernetics Complexity theory		Postmodern systems thinking

Table 1.3 contrasts methodologies and problems (Jackson & Keys, 1984). Again, only one cell of the table includes traditional OR and systems analysis. For other than mechanical problems with a single decision maker, much more participative approaches are warranted, at least if the goal is solving the problem of interest rather than just modeling the "physics" of the context.

Table 1.4 summarizes Ulrich's (1988) levels of system practice. He differentiates hard versus soft in terms of three categories – one hard and two versions of soft. One class of soft management addresses change while the other addresses conflict. The key disciplines and tools vary substantially across these three categories.

TABLE 1.3   Methodologies versus Problems (Jackson & Keys, 1984)

	Mechanical	Systemic
Unitary – one decision maker	Operations research Systems engineering Systems analysis	Organizational cybernetics socio-technical systems
Pluralist – multiple independent decision makers	Singerian inquiry systems Strategic assumption methods Wicked problem formulations	General systems theory Complex adaptive systems Soft systems methodology

Table 1.5 summarizes Jackson's (2003) Critical Systems Practice. The most important aspect of his guidance is to remain open to the range of possibilities in Tables 1.1–1.4. From the perspective of understanding complex systems, this means that the nature of models entertained should be driven by the issues of interest, the phenomena underlying these issues, and the orientations of the key stakeholders in the problem framing and solving processes.

Pidd (2004) offers the notion of complementarity as a way of rationalizing the relationship between hard and soft approaches. He argues that hard and soft approaches are complementary to each other, but their complementarity is asymmetric. He asserts that any problem situation in human affairs will always at some level entail differences in world views that the "soft" approaches can be used to explore. Within that exploration, any or all of the hard approaches can be adopted as a conscious strategy. The reverse strategy is not available because it entails abandoning the ontological stance of hard approaches. In other words, hard approaches are often inextricably tied to paradigms and assumptions that are central to their problem-solving power.

There is a wealth of formal methods that can play a role in systems practice. Approaches to systems modeling, from a range of disciplinary perspectives, are discussed in Haskins (2006) and in Sage and Rouse (2009). A variety of paradigm-specific treatments are also available, such as Forrester's (1961) classic on systems dynamics modeling and Sterman's (2000) contemporary treatment of system dynamics modeling. Chapter 9 elaborates a variety of formal theories in terms of typical assumptions and outcomes predicted, along with brief expositions of the basic mathematics.

Gharajedaghi (2011) articulates a system methodology for supporting complex adaptive systems. The methodology focuses on functions, structure,

TABLE 1.4 Levels of Systems Practice (Ulrich, 1988)

Aspect	Operational Systems Management	Strategic Systems Management	Normative Systems Management
Dominating interpretation	Systematic	Systemic	Critical idea of reason
Strand of systems thinking	Hard – mechanistic paradigm	Soft – evolutionary paradigm	Soft – normative paradigm
Dimension of rationalization	Instrumental	Strategic	Communicative
Main object of rationalization	Resources – means of production	Policies – steering principles	Norms – collective preferences
Task of the expert	Management of scarceness	Management of complexity	Management of conflict
Type of pressure	Costs	Change	Conflict
Basic approach	Optimization	Steering capacity	Consensus
Goodness criterion	Efficient	Effective	Ethical
Theory-practice mediation	Decisionistic	Technocratic	Pragmatistic
Key disciplines	Decision theory, economics, engineering	Game theory, ecology, social sciences	Discourse theory, ethics, critical theory
Example tools	Cost–benefit analysis, linear optimization	Sensitivity analysis, large-scale simulation	Systems assessment, ideal planning
Trap to avoid	Suboptimization	Social technology	Excluding the affected

TABLE 1.5   Critical Systems Practice (Jackson, 2003)

Creativity	
Task	To highlight significant concerns, issues, and problems
Tools	Creativity-enhancing devices employing multiple perspectives
Outcome	Dominant and dependent concerns, issues, and problems
Choice	
Task	To choose an appropriate generic systems methodology
Tools	Methods for revealing methodological strengths and weaknesses
Outcome	Dominant and dependent generic systems methodologies
Implementation	
Task	To arrive at and implement specific positive change proposals
Tools	Generic systems methodologies
Outcome	Highly relevant and coordinated change yielding improvements
Reflection	
Task	To produce learning about the problem and solution
Tools	Clear understanding about the current state of knowledge
Outcome	Research findings that fed back into practice

and processes. To define functions, he argues that one should clarify which products solve which problems for which customers. To define structure, he advances the idea of a modular design that defines complementary relationships among relatively autonomous units. Finally, design of processes involves using a multidimensional modular design based on the triplet input (technology), output (products), and environments (markets).

This brief discussion of systems approaches serves to set the stage for alternative approaches to understanding complex systems and enterprises. The nature of these systems usually precludes fully modeling them with first-principles physics models. These systems are, by no means, as mechanistic and predictable as purely physical systems like bouncing balls or gear trains. Yet, there are well-developed approaches for addressing problem solving in complex systems and enterprises. Valid predictions, and occasionally optimization, are certainly of interest. However, insights into phenomena, sensitivities to key parameters, and consensus building are often the overarching goals.

The material discussed in this section sets the stage for the methodological discussions in Chapter 2. The emphasis on problem formulation and hard versus soft approaches are highly relevant. The first four steps of our ten-step methodology are very much focused on problem formulation. The great emphasis placed on visualization first and computation later enables taking advantage of "soft" approaches early and only resorting to "hard" approaches for aspects of problems that warrant such investments.

PHENOMENA AS THE STARTING POINT

The construct of "phenomena" is central to this book. Problem solving should not begin with the selection of mathematical or computational models, but instead should commence with consideration of the phenomena that must be understood to successfully answer the questions that motivated the modeling effort in the first place.

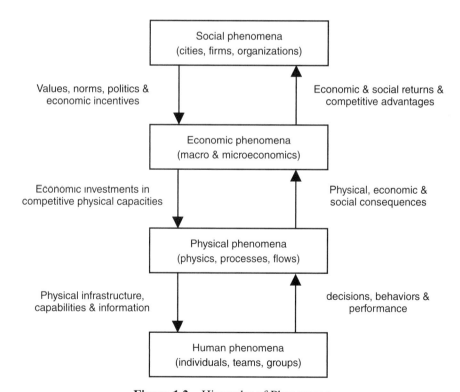

Figure 1.2 Hierarchy of Phenomena

TABLE 1.6   Eight Classes of Phenomena

Class of Phenomena	Example Phenomena of Interest
Physical, natural	Temporal and spatial relationships and responses
Physical, designed	Input–output relationships, responses, stability
Human, individuals	Task behaviors and performance, mental models
Human, teams or groups	Team and group behavior and performance
Economic, micro	Consumer value, pricing, production economics
Economic, macro	Gross production, employment, inflation, taxation
Social, organizational	Structures, roles, information, resources
Social, societal	Castes, constituencies, coalitions, negotiations

Figure 1.2 provides a framework for thinking about phenomena and relationships among phenomena. There are four levels – physical, human, economic, and social – as well as typical relationships among phenomena. Of course, there are many subtleties that are not reflected in Figure 1.2, but will be elaborated in Chapters 3–7.

Table 1.6 provides a glimpse into the eight classes of phenomena addressed in this book. The overall taxonomy of phenomena is elaborated in Chapter 3, while physical, human, economic, and social phenomena are addressed in Chapters 4–7, respectively. We also discuss in Chapter 3 the phenomena associated with the six archetypal problems that are introduced in Chapter 2.

OVEVIEW OF BOOK

This final section of this introductory chapter provides synopses of the chapters in this book and the lines of reasoning that connect them.

Chapter 1: Introduction and Overview

This chapter begins by placing modeling and visualization of complex systems in the context of the evolution of the systems movement, its philosophical background, and a wide range of seminal concepts. Constructs associated with complexity and complex systems are discussed. The important contrast between complex and complicated systems is elaborated. The last century of systems practice is briefly reviewed to provide a foundation for the methodology advocated in this book. The use of phenomena as a starting point is then argued. Finally, an overview of the book is provided.

Chapter 2: Overall Methodology

This chapter begins with a discussion of human-centered methods and tools. The emphasis is on assuring that a methodology is both useful and usable. We then discuss six problems archetypes that are addressed throughout this book. The choice of this set of problems was motivated by the desire to assure that the methodology not be problem specific. Attention then turns to the overall ten-step methodology. An example is used to illustrate application of the methodology to congestion pricing of urban traffic. The chapter concludes with discussion of an environment that supports use of the methodology.

Chapter 3: Perspectives on Phenomena

In this chapter, we explore the fundamental nature of phenomena and the role that this construct plays in technology development and innovation. The construct is discussed from both historical and contemporary perspectives. Numerous historical and contemporary examples are used to illustrate the evolution of technology as well as the increasing scope of its application. A taxonomy of phenomena is introduced, with particular attention paid to its behavioral and social components. Use of the taxonomy is illustrated using the six archetypal problems introduced in Chapter 2. Finally, visualization of phenomena is discussed in the context of the examples used throughout Chapters 4–7.

Chapter 4: Physical Phenomena

This chapter considers two types of physical phenomena. First, we discuss naturally occurring phenomena such as weather and water flow. We consider two examples – human biology and urban oceanography. Then we address designed phenomena such as systems engineered to move people and goods. Examples of interest here include vehicle powertrains and manufacturing processes. We also discuss the intersection of designed and natural physical phenomena, which is a central issue in urban oceanography. The chapter concludes with an elaboration of the archetypal example of deterring or identifying counterfeit parts.

Chapter 5: Human Phenomena

This chapter begins by contrasting descriptive and prescriptive approaches. Descriptive approaches focus on data from past instances of the phenomena of interest. Prescriptive approaches attempt to calculate what humans should do given the constraints within which they have to operate. A wide range of models of human behavior and performance are outlined. Examples discussed include manual control, problem solving, and multitask decision making. This exposition is followed by a detailed look at the traffic control problem from our set of archetypal problems. Many of the models discussed make assumptions about what humans know relative to the tasks at hand. Some of this knowledge is characterized using the notion of "mental models." The nature of this construct is discussed in terms of approaches to assessing mental models and use of the outcomes of such assessments. Finally, fundamental limits in modeling human behavior and performance are addressed.

Chapter 6: Economic Phenomena

This chapter reviews concepts and models from microeconomics, macroeconomics, and behavioral economics. Within microeconomics, the theory of the firm and the theory of the market are reviewed. Examples presented include optimal pricing and the economics of investing in people. Macroeconomic issues frame the context for microeconomic decisions. We consider gross domestic product growth, tax rates, interest rates, and inflation, which can strongly affect the economic worth of alternative investments. Behavioral economics is addressed in terms of how people actually behave rather than how traditional economists assume they behave. A final section addresses the economics of health-care delivery. This example is laced with the microeconomic and macroeconomic phenomena discussed throughout this chapter.

Chapter 7: Social Phenomena

This chapter begins with consideration of the social phenomena identified for the six archetypal problems discussed throughout this book. Emergent and designed phenomena are then contrasted, as well as direct versus representative political systems. The overall problem of modeling complex social systems is then considered, with an example of modeling the earth as a system. Several approaches to modeling social systems are presented, including physics-based formulations, network theory, game theory, and simulations. Examples include castes and outcastes, acquisition as a game, port and airport evacuation, and the emergence of cities. Attention then shifts to urban resilience, including introduction of a framework for understanding the full nature of resilience.

Chapter 8: Visualization of Phenomena

This chapter first briefly addresses human vision as a phenomenon, primarily to recognize the topic as important but also to move beyond the science of vision to the design of visualizations. We next review the basics of visualization to provide grounding for the subsequent discussions. We then address the purposes of visualization. The object of design, it is argued, is the fulfillment of purposes. A visualization design methodology is then presented and illustrated with an example from helicopter maintenance. Visualization tools are then briefly reviewed. Various case studies from Stevens Institute's *Immersion Lab* are discussed. The notion of policy flight simulators is introduced and elaborated. Finally, results are presented from an extensive study of what users want from visualizations and supporting computational tools.

Chapter 9: Computational Methods and Tools

This chapter addresses modeling paradigms potentially useful for addressing the phenomena associated with the six archetypal problems discussed throughout the book. Paradigms discussed include dynamic systems theory, control theory, estimation theory, queuing theory, network theory, decision theory, problem solving theory, and finance theory. A multilevel modeling framework is used to illustrate how the different modeling paradigms can be employed to represent different levels of abstraction and aggregation of an overall problem. The next consideration is moving from representation to computation. This includes discussion of model composition and issues of entangled states and consistency of assumptions. This chapter also provides a brief overview of software tools available to support use of these computational approaches.

Chapter 10: Perspectives on Problem Solving

This chapter brings all the material in this book together to discuss a large set of case studies involving well over 100 enterprises and several thousand participants. Case studies cover broad areas of business planning, new product planning, technology investments, and enterprise transformation. Discussion of these case studies focuses on how problem solving was addressed, the roles that interactive models played in problem solving, and the types of insights and decisions that resulted. These observations on problem solving are summarized in terms of guidance on starting assumptions, framing problems, and implementing solutions. This chapter concludes with consideration of key research issues that need to be addressed to advance the approach to problem solving advocated in this book.

REFERENCES

Ackoff, R.L. (1971). Toward a system of systems concepts. *Management Science* 17 (11), 661–671.

Ackoff, R.L. & Emery, F. (1972). *On Purposeful Systems: An Interdisciplinary Analysis of Individual and Social Behavior as a System of Purposeful Events*. Chicago: Aldine-Atherton.

Ackoff, R.L., Churchman, C.W., & Arnoff, E.L. (1957). *Introduction to Operations Research*. New York: John Wiley & Sons.

Adams, K.M., Hester, P.T., Bradley, J.M., Meyers, T.J., & Keating, C.B. (2014). Systems theory as the foundation for understanding systems. *Systems Engineering*, 17 (1), 112–123.

Alderson, D.L., & Doyle, J.C. (2010). Contrasting views of complexity and their implications for network-centric infrastructures. *IEEE Transaction on Systems, Man and Cybernetics – Part A: Systems and Humans*, 40 (4), 839–852.

Arrow, K. J., (1951). *Social Choice and Individual Values*. New York: Wiley.

Arrow, K. J., & Debreu, G. (1954). Existence of a competitive equilibrium for a competitive economy. *Econometrica*, 22 (3), 265–90.

Ashby, W.R. (1952). *Design for a Brain*. London: Chapman & Hall.

Ashby, W. R. (1956). *An Introduction to Cybernetics*. London: Chapman & Hall.

Baldwin, W.C., Sauser, B., Boardman, J., & John, L. (2010). A Typology of Systems Paradoxes. *Information Knowledge Systems Management*, 9 (1), 1–15.

Bellman, R.E. (1957). *Dynamic Programming*. Princeton, NJ: Princeton University Press.

Bertalanffy, K.L.V. (1968). *General System theory: Foundations, Development, Applications*. New York: George Braziller.

Chaitin, G.J. (1974). Information-theoretic limitations of formal systems. *Journal of ACM*, 21 (3), 409–424.

Checkland, P. (1993). *Systems Thinking, Systems Practices*. Chichester, UK: John Wiley.

Churchman, C.W. (1971). *The Design of Inquiring Systems, Basic Concepts of Systems and Organizations*. New York: Basic Books.

Emery, F. & Trist, E. (1965). The causal texture of organizational environments. *Human Relations*, 18, 21–32.

Emery, F. & Trist, E. (1973). *Toward a Social Ecology*. London: Plenum Press,

Forrester, J. W. (1961). *Industrial Dynamics*. Westwood, MA: Pegasus Communications.

Gharajedaghi, J. (2011). *Systems Thinking: Managing Chaos and Complexity: A Platform for Designing Business Architecture* (3rd Edition). New York: Morgan Kaufmann.

Haskins, C. (2006). *Systems Engineering Handbook*. San Diego, CA: International Council on Systems Engineering.

Jackson, M.C. (2003). *Systems Thinking: Creative Holism for Managers*. Chichester, UK: John Wiley.

Jackson, M.C., & Keys, P. (1984). Towards a System of Systems Methodologies. *Journal of the Operational Research Society*, 35 (6), 473–486.

Kahneman, D. (2011). *Thinking, Fast and Slow*. New York: Farrar, Straus and Giroux.

Kalman, R.E. (1960). A new approach to linear filtering and prediction problems. *Transactions of the ASME, Journal of Basic Engineering*, 82, 34–45.

Meadows, D.H. (2008). *Thinking in Systems: A Primer*. White River Junction, VT: Chelsea Green.

Miller, G.A. (1956). The magic number seven, plus or minus two: Some limits on our capacity for processing information. *Psychological Review*, 63 (2), 81–97.

Morse, P.M. (1958). *Queues, Inventories, and Maintenance*. New York: John Wiley & Sons.

Morse, P.M. (1968). *Library Effectiveness*. Cambridge, MA: MIT Press.

Morse, P.M., & Kimball, G.E. (1951). *Methods of Operations Research*. New York: John Wiley & Sons.

Parsons, T. (1937). *The Structure of Social Action*. New York: McGraw-Hill

Parsons, T. (1951a). *The Social System*. Glencoe, IL: The Free Press.

Parsons, T. (1951b). *Toward a General Theory of Action*. Cambridge, MA: Harvard University Press.

Parsons, T., & Smelser, N.J. (1956). *Economy and Society: A Study in the Integration of Economic and Social Theory*. London: Routledge & Kegan Paul.

Pidd, M. (Ed.). (2004). *Systems Modeling: Theory and Practice*. Chichester, UK: John Wiley.

Poli, R. (2013). A note on the difference between complicated and complex social systems. *Cadmus*, 2 (1), 142–147.

Pontryagin, L. S., Boltyanskii, V. G., Gamkrelidze, R. V., and Mishchenko, E. F. (1962). *A Mathematical Theory of Optimal Processes*. New York: Gordon & Breach.

Rittel, H.W.J., & Webber, M.M. (1973). Dilemmas in a general theory of planning. *Policy Science*, 4, 155–169.

Robinson, S., Brooks, R., Kotiadis, K., & van der Zee, D-J. (2011). *Conceptual Modeling for Discrete-Event Simulation*. Boca Raton, FL: CRC Press.

Rouse, W.B. (2000). Managing complexity: Disease control as a complex adaptive system. *Information Knowledge Systems Management*, 2 (2), 143–165.

Rouse, W.B. (2003). Engineering complex systems: Implications for research in systems engineering. *IEEE Transactions on Systems, Man, and Cybernetics – Part C*, 33 (2), 154–156.

Rouse, W.B. (2005). *Thoughts on Complex Systems: An Initial Framework and Discussion Questions*. Atlanta, GA: Tennenbaum Institute, October.

Rouse, W.B. (2007). Complex engineered, organizational & natural systems: Issues underlying the complexity of systems and fundamental research needed to address these issues. *Systems Engineering*, 10 (3), 260–271.

Rouse, W.B. (2008). Healthcare as a complex adaptive system: Implications for design and management. *The Bridge*, 38 (1), 17–25.

Rouse, W.B., & Hammer, J.M. (1991). Assessing the impact of modeling limits on intelligent systems. *IEEE Transactions on Systems, Man, and Cybernetics*, SMC-21(6), 1549–1559.

Rouse, W.B., Hammer, J.M., & Lewis, C.M. (1989). On capturing humans' skills and knowledge: Algorithmic approaches to model identification. *IEEE Transactions on Systems, Man, and Cybernetics*, SMC-19(3), 558–573.

Rouse, W.B., Boff, K.R., & Sanderson, P. (Eds.).(2012). *Complex Socio-Technical Systems: Understanding and Influencing Causality of Change*. Amsterdam: IOS Press.

Rouse, W.B., & Morris, N.M. (1986). On looking into the black box: Prospects and limits in the search for mental models. *Psychological Bulletin*, 100(3), 349–363.

Rouse, W.B., & Serban, N. (2011). Understanding change in complex socio-technical systems: An exploration of causality, complexity and modeling. *Information Knowledge Systems Management*, 10, 25–49.

Sage, A.P., & Rouse, W.B. (Eds.).(2009). *Handbook of Systems Engineering and Management*. Hoboken, NJ: Wiley.

Shannon, C. (1948). A mathematical theory of communication. *The Bell System Technical Journal*, 27, 379–443, 623–656.

Simon, H.A. (1957). *Models of Man*. New York: John Wiley.

Simon, H.A. (1962). The architecture of complexity. *Proceedings of the American Philosophical Society*, 106 (6), 467–482.

Snowden, D.J., & Boone, M.E. (2007). A leader's framework for decision making. *Harvard Business Review*, November, 69–76.

Sterman, J.D (2000). *Business Dynamics: Systems Thinking and Modeling for a Complex World*. New York: McGraw-Hill.

Tolk, A. (Ed.).(2013). *Ontology, Epistemology, and Teleology for Modeling and Simulation*. Berlin Heidelberg: Springer-Verlag.

Trist, E., & Bamforth, W. (1951). Some social and psychological consequences of the long wall method of coal getting. *Human Relations*, 4, 3–38.

Ulrich, W. (1988). Systems thinking, systems practice, and practical philosophy: A program of research. *Systems Practice*, 1 (2), 137–163.

Weinberg, G.M. (2001). *An Introduction to General Systems Thinking*. London: Dorset House.

Wiener, N. (1948, 1961). *Cybernetics: Or Control and Communication in the Animal and the Machine*. Cambridge, MA: MIT Press.

2

OVERALL METHODOLOGY

INTRODUCTION

I have been researching and developing methods and tools for almost 50 years. This typically involves generalizing something you have learned to do well yourself into something others can do successfully. This is not as easy as it might sound. My experience has been that the vast majority of methods and tools are only used well by their originators.

Several methods and tools that I pursued were intended to help users pursue human-centered design, a process of considering and balancing the concerns, values, and perceptions of all the stakeholders in a design (Rouse, 1991, 2007). By stakeholders, I mean users, customers, developers, maintainers, and competitors. The premise of human-centered design is that all stakeholders need to perceive methods and tools to be valid, acceptable, and viable.

Valid methods and tools help solve the problems for which they are intended. Acceptable methods and tools solve problems in ways that stakeholders prefer. Viable methods and tools provide benefits that are worth the costs of use. Costs here include the efforts needed to learn and use methods and tools, not just the purchase price.

Modeling and Visualization of Complex Systems and Enterprises:
Explorations of Physical, Human, Economic, and Social Phenomena, First Edition. William B. Rouse.
© 2015 John Wiley & Sons, Inc. Published 2015 by John Wiley & Sons, Inc.

The methodology presented in this chapter is intended to increase validity, acceptability, and viability beyond that usually experienced with the ways in which problems of the scope addressed in this book are usually pursued. This begs the question of what shortcomings plague existing approaches.

First and foremost are viability issues. Sponsors of modeling efforts complain that they take too long and are too expensive. This is due in part to the business processes of sponsors. However, more fundamentally, much time and money go into developing aspects of models that, at least in retrospect, were not needed to address the questions of interest.

Second are acceptability issues. Many key stakeholders in the types of problems addressed in this book are not educated in modeling and simulation. Nevertheless, they are often highly talented, have considerable influence, and will not accept that the optimal solution, somehow magically produced, is $X = 12$. We need methods and tools that are more engaging for these types of stakeholders (Rouse, 1998, 2014).

Third are validity issues. There is often concern that overall models composed of several legacy models are of questionable validity (Pennock & Rouse, 2014a, 2014b). This concern is due in part to the possibility that assumptions are inconsistent across component models. There is also the issue of coupled states across component models, which can lead to incomputable or unstable computations. This is particularly plaguing when one does not know it is happening.

The methodology presented in this chapter overcomes these issues in several ways. First, we avoid overmodeling by delaying equations and deep computation until the fifth step of the 10-step methodology. The first four steps of the methodology focus on problem formulation, with particular emphasis on interactive pruning of the problem space prior to any in-depth explorations.

Second, we have found that key stakeholders value being immersed in interactive visualizations of the phenomena, and relationships among phenomena associated with their domain and the questions of interest. This enables them to manipulate controls and explore responses. This is typically done in a group setting with much discussion and debate.

Third, two steps of the methodology explicitly address model composition issues, including agreeing on a consistent set of assumptions across models. This prompts delving into the underpinnings of each component model. The concern is much more fundamental than whether two software codes can syntactically communicate. The overarching question is whether connecting two models will yield results that are valid in the context of the questions at hand.

This chapter has begun with a discussion of human-centered methods and tools and the issues of viability, acceptability, and validity. We next discuss six problems archetypes that are addressed throughout this book. These problems provide the use cases for the methodology. Attention then turns to the overall 10-step methodology. An example use of the methodology is discussed. The chapter concludes with consideration of how use of the methodology can be supported.

PROBLEM ARCHETYPES

It is very difficult to discuss complex systems in general. A major difficulty is that context matters. In the absence of context, the discussion is, at best, rather abstract. This section elaborates six examples of complex systems that were carefully chosen to stretch the overall intellectual framework presented in this book, as well as support identification of a range of phenomena of interest across this variety of problem areas. These examples evolved from earlier versions discussed by Pennock and Rouse (2014a, b).

We first discuss deterring or identifying counterfeit parts in aerospace and defense systems. In this case, the systems of interest were engineered, as was the organizational system for procuring these systems. The second example concerns financial systems and the bursting of bubbles. The investment products of interest were engineered or designed, as was the context of investment, although the design of these products and context was not thought to be an example of engineering.

The next two examples focus on cities. First, we consider human responses to urban threats (e.g., hurricanes) and urban resilience. Then, we focus on one specific urban system in the context of traffic control via congestion pricing. In both cases, we have engineered networks of urban infrastructure embedded in the complex behavioral and social contexts of contemporary cities. For these examples, much less is designed in a formal sense. Many phenomena are emergent.

The final two examples address healthcare. First, we address the impacts of investments in healthcare delivery and how payment schemes affect investments and consequent health outcomes. We then address a particular health threat - human biology and cancer. This is a complex system in that the biological system succeeds or fails in the context of human lifestyles and environmental risks and consequences. Overall, these examples range from broad socioeconomic systems to individual humans functioning in a broader context.

Deterring or Identifying Counterfeit Parts

Thousands of suppliers provide millions of parts that flow through supply chains to subsystem assembly and then final assembly of the overall system.

Performance and reliability of these parts determine performance and availability of the overall system to serve its intended purpose, for example, transportation or defense. Downward pressures on suppliers' pricing of parts potentially undermine suppliers' profit margins, motivating them to cut costs somewhere. Leaning of materials and production costs reaches diminishing returns for one or more suppliers, which causes them to intentionally decrease quality of parts. Counterfeit parts are detected by increased and tightened inspection and/or inhibited by economic incentives for suppliers, both of which exacerbate cost problems. Phenomena of interest in this example include physical phenomena of discrete flow and system performance, microeconomic phenomena of costs, quality and process operations, and macroeconomic phenomena of costs, prices, and profits.

Financial Systems and Bursting Bubbles

Demand for high-quality investments exceeds available supply. The financial sector is incentivized to increase supply by selling low-quality investments. Financial engineers create new derivatives that combine previously high-risk investments in a way that is intended to reduce the overall risk based on the assumption of low correlation among assets. These derivatives are sold as high-quality investments. This lowers the cost of capital to previously low-quality investments. This results in over-investment in low-quality assets, which increases prices and consequently returns on investment for asset holders. Demand for additional low-quality assets increases. As the demand for low-quality assets exceeds supply, suppliers lower minimum standards and/or create fraudulent assets. The resulting positive feedback loop creates an asset bubble that introduces a systemic risk that increases the underlying risk of the "high-quality" derivatives, that is, the bubble increases correlation among the assets. Eventually, the lowest quality assets begin to default, causing a chain reaction resulting in a crash of financial markets. Phenomena of interest in this example include macroeconomic phenomena of demand, supply, products, capital, investment, assets, prices, sales, profits, risks, and crashes.

Human Responses and Urban Resilience

A projected storm surge leads to predictions of flooding within a specific urban topography. Projected flooding leads to anticipated deterioration of infrastructure for transportation, energy, and so on. Projections are communicated to inhabitants and subsequently communicated among inhabitants, resulting in altered perceptions. Perceptions (and later experiences) of impending deterioration lead people to adapt by planning to move to higher ground or to leave the area. Plans are shared among inhabitants, resulting in altered intentions. Intentions to move or leave enable projections of demands on urban infrastructure. Projections result in altered communications to inhabitants as well as among inhabitants. The results can range from resilient responses to complete gridlock. Phenomena of interest in this example include physical phenomena of continuous flow and system performance, human phenomena of communications, decision making and social interaction, and microeconomic phenomena of process operations.

Traffic Control via Congestion Pricing

Congestion in particular urban areas causes increased transit times in these areas. Time-varying time-unit pricing is adopted for use of these roads. Government likes the revenue. Business in these areas may be concerned about loss of traffic. Motorists respond by avoiding these areas and using other roads or modes of transportation. Increasing demand for alternatives affects congestion in these areas. Motorists communicate with each other in search of shortcuts and avoiding tolls. Thus, flows affect pricing, and pricing affects flows, with no guarantee of equilibrium. Phenomena of interest in this example include physical phenomena of continuous flow, human phenomena of decision making and social interaction, and macroeconomic phenomena of prices and revenue.

Impacts of Investments in Healthcare Delivery

Demand for services (e.g., chronic disease care) and payment models (e.g., by Medicare) drives investments in capacities to provide services by healthcare providers. Capacities in the form of people, equipment, and facilities are scheduled to meet demands. Use of capacities as scheduled results in outcomes, costs, and revenue. Quality of outcomes results in decreased demands for some services (e.g., reduced Emergency Department visits and in-patient admissions), but increased demands for others (e.g., out-patient chronic disease management). More subtly, decreased capacities to care for diseases with low payments can cause increased prevalence of other diseases – for instance, poor care for early diabetes mellitus leads to increases in coronary heart disease. Phenomena of interest in this example include microeconomic phenomena of process operations and macroeconomic phenomena of prices, demand, and supply.

Human Biology and Cancer

Human genes express proteins that result in 50 trillion cells, with several hundred distinct types, which compose tissues that, in turn, compose organs, muscles, and so on within cardiovascular, pulmonary, vestibular, and other systems. The nervous system, a network of specialized cells, coordinates the actions of humans and sends signals from one part of its body to another. These cells send signals either as electrochemical waves traveling along thin fibers called axons or as chemicals released onto other cells. Signaling aberrations result in dysfunctions in the control of cellular processes, for example, cell growth and death, resulting in phenomena such as cancer. Targeted therapies, for example, signal transduction inhibitors, can be used to treat cancers that result from aberrations to signaling pathways involved in cell growth. Cancers evolve in how they react to therapies. Phenomena of interest in this example include human phenomena of normal biology, abnormal biology, and decision making.

Comparison of Problems

These six examples have several common characteristics:

- All involve behavioral and/or social phenomena, directly or indirectly;
- All involve effects of human variability, both random and systemic;
- All involve economics (pricing) or financial consequences;
- All include both designed (engineered) and emergent aspects.

There are also important distinctions:

- Counterfeit Parts and Financial System involve deception by a subset of the actors;
- Healthcare Delivery and Human Biology involve aberrant functioning by a subset of the actors;
- Congestion Pricing and Urban Resilience involve aggregate consequences (e.g., traffic) of all actors.

Another important distinction is between two classes of problems.

- *Bottom-up*. Detection and remediation of aberrant actors involves stratifying actors and exploring behaviors of each stratum in different ways
  - Aberrant actors tend to react to remediation strategies, eventually undermining their effectiveness.

- *Top-down.* Economic strategies, for example, pricing, payment models, procurement practices, based on aggregate behaviors
  - Individual actors tend to react to aggregate strategies, often undermining the desired consequences.

Considering how the phenomena associated with these examples might be represented, three common features should be noted. First, the set of phenomena associated with a problem can be represented at different levels of abstraction, for example, individual instances of counterfeiting versus macroeconomic policies that motivate counterfeiting. Second, each example has phenomena of interest that emerge within each layer of abstraction. This would suggest that a different representation of the complex system would be relevant for each layer. Third, each example exhibits feedback loops that cut across two or more layers. For example, the incentive to counterfeit increases with declining supplier profit margins. High-level policies designed to combat counterfeiting could raise costs at the lower levels. This could further erode profit margins and actually increase the incentive to counterfeit. Thus, the counterfeiting problem cannot be addressed without considering the relationships between the different layers of the complex system. We discuss the phenomena associated with these six problems in Chapter 3.

METHODOLOGY

Experience has shown that models should be developed with a clear intent, with defined scope and givens. Initial emphasis should be on alternative views of phenomena important to addressing the questions of interest. Selected views can then be more formally modeled and simulated. The following 10-step methodology provides a structure to support this approach to modeling.

Step 1: Decide on the central questions of interest. The history of modeling and simulation is littered with failures of attempts to develop models without clear intentions in mind. Models provide means to answer questions. Efforts to model socio-technical systems are often motivated by decision makers' questions about the feasibility and efficacy of decisions on policy, strategy, operations, and so on. The first step is to discuss the questions of interest with the decision maker(s), define what they need to know to feel that the questions are answered, and agree on key variables of interest.

Step 2: Define key phenomena underlying these questions. The next step involves defining the key phenomena that underlie the variables associated with the questions of interest. Phenomena can range from physical, behavioral, or organizational, to economic, social, or political. Broad classes of phenomena across these domains include continuous and discrete flows, manual and automatic control, resource allocation, and individual and collective choice. Mature domains often have developed standard descriptions of relevant phenomena.

Step 3: Develop one or more visualizations of relationships among phenomena. Phenomena can often be described in terms of inputs, processes, and outputs. Often the inputs of one phenomenon are the outputs of other phenomena. Common variables among phenomena provide a basis for visualization of the set of key phenomena. Common visualizations methods include block diagrams, IDEF, influence diagrams, and systemigrams.

Step 4: Determine key trade-offs that appear to warrant deeper exploration. The visualizations resulting from step 3 often provide the basis for in-depth discussions and debates among members of the modeling team as well as the sponsors of the effort, which hopefully includes the decision makers who intend to use the results of the modeling effort to inform their decisions. Lines of reasoning, perhaps only qualitative, are often verbalized that provides the means for immediate resolution of some issues, as well as dismissal of some issues that no longer seem to matter. New issues may, of course, also arise.

Step 5: Identify alternative representations of these phenomena. Computational representations are needed for those phenomena that will be explored in more depth. These representations include equations, curves, surfaces, process models, agent models, and so on – in general, instantiations of standard representations. Boundary conditions can affect choices of representations. This requires deciding on fixed and variable boundary conditions such as GDP growth, inflation, carbon emissions, and so on. Fixed conditions can be embedded in representations while variable conditions require controls such as slider bars to accommodate variations – see step 9.

Step 6: Assess the ability to connect alternative representations. Representations of phenomena associated with trade-offs to be addressed in more depth usually require inputs from other representations and produce outputs required by other representations. Representations may differ in terms of dichotomies such as linear versus nonlinear, static versus dynamic, deterministic versus stochastic, continuous versus discrete, and so on. They may also differ in terms of basic assumptions, for example, Markov versus non-Markovian processes. This step involves determining what can be meaningfully connected together.

Step 7: Determine a consistent set of assumptions. The set of assumptions associated with the representations that are to be computationally connected need to be consistent for the results of these computations to be meaningful.

At the very least, this involves synchronizing time across representations, standardizing variable definitions and units of measures, and agreeing on a common coordinate system or appropriate transformations among differing coordinate systems. It also involves dealing consistently with continuity, conservation, and independence assumptions.

Step 8: Identify data sets to support parameterization. The set of representations chosen and refined in steps 5–7 will have parameters such as transition probabilities, time constants, and decay rates that have to be estimated using data from the domain(s) in which the questions of interest are to be addressed. Data sources need to be identified and conditions under which these data were collected determined. Estimation methods need to be chosen, and in some cases developed, to provide unbiased estimates of model parameters.

Step 9: Program and verify computational instantiations. To the extent possible, this step is best accomplished with commercially available software tools. The prototyping and debugging capabilities of such tools are often well worth the price. A variant of this proposal is to use commercial tools to prototype and refine the overall model. Once the design of the model is fixed, one can then develop custom software for production runs. The versions in the commercial tools can then be used to verify the custom code. This step also involves instantiating interactive visualizations with graphs, charts, sliders, radio buttons, and so on.

Step 10: Validate model predictions, at least against baseline data. The last step involves validating the resulting model. This can be difficult when the model has been designed to explore policies, strategies, and so on for which there inherently is no empirical data. A weak form of validation is possible by using the model to predict current performance with the "as is" policies, strategies, and so on. In general, models used to explore "what if" possibilities are best employed to gain insights that can be used to frame propositions for subsequent empirical study.

Summary

The logic of the 10-step methodology can be summarized as follows, with emphasis on steps 1–7:

- Define the question(s) of interest
- Identify relevant phenomena
- Visually compose phenomena
- Identify useful representations
- Computationally compose representations

Note that this logic places great emphasis on problem framing and formulation. Deep computation is preserved for visually identified critical trade-offs rather than the whole problem formulation. Steps 8–10 of the methodology are common to many methodologies.

Not all problems require full use of this 10-step methodology. Often visual portrayals of phenomena and relationships are sufficient to provide the insights of interest. As just noted, such views are also valuable for determining which aspects of the problem should be explored more deeply.

AN EXAMPLE

The purpose of this example is to illustrate the line of reasoning typically associated with use of the 10-step methodology outlined earlier. Specific representations chosen and computational results are not presented for this notional example, as our intent is focused solely on how to think about modeling complex systems using the proposed methodology.

Step 1: Decide on the central questions of interest. The questions of interest concern the design of congestion pricing models to discourage personal use of cars in the central city during times of potential peak demand. This involves the design of the magnitude and timing of congestion pricing, including the implication of alternative designs.

Step 2: Define key phenomena underlying these questions. The phenomena of potential interest initially include:

- *Traffic control*: Magnitude and timing of prices per unit time for use of targeted inner city zone(s).
- *Traffic congestion*: Traffic densities, delays, and accidents in these zones at different time of the day with and without congestion pricing.
- *Vehicle control*: Driver–vehicle performance as a function of varying levels of congestion.
- *Vehicle performance*: Acceleration and braking performance in varying levels of stop-and-go driving.
- *Powertrain performance*: Energy efficiency and pollution emissions as a function of vehicle performance and control as a function of varying levels of congestion.

Step 3: Develop one or more visualizations of relationships among phenomena. Development of a visualization of relationships among this set of phenomena leads to two important conclusions. First, powertrain and vehicle performance are key to predicting the likely energy consumption and pollution emission outcomes of a congestion pricing model, but are not central to designing the model. Thus, these two phenomena can be shelved, at least until energy and pollution issues are of interest.

The second conclusion is that two important phenomena are missing. One is drivers' decision making in response to congestion pricing. More specifically, what percent of drivers choose to avoid the central city as a function of the pricing level? Second, what alternative modes of transportation do drivers who chose to avoid the central city chose instead? Do they avoid traveling, or take public transportation, or drive via different routes. Thus, drivers' decision making is a critical phenomenon.

The relationships among key phenomena are shown in Figure 2.1. Drivers make decisions about routes based on perceptions of traffic conditions and posted prices – recall they are intended to vary in time. We are concerned with the broad road network because drivers' decisions can lead to congestion on roads other than those with congestion pricing. Traffic management adjusts its pricing model according to its traffic predictions. Traffic management's degrees of freedom are constrained by laws and regulations. Drivers' decisions in response to the posted prices driven by this model result in levels of traffic efficiency and safety across the broad network. Drivers' behaviors are also influenced by social values and norms. The resulting levels of traffic efficiency and safety strongly influence public opinion.

Step 4: Determine key trade-offs that appear to warrant deeper exploration. The key issue was first seen as determining what level of pricing would discourage driving in the central city during high-demand periods. With a little reflection, it is obvious that exorbitant prices will discourage everyone. But this will eliminate all traffic and generate no revenue for the city. Inner city merchants certainly do not want to eliminate all traffic and city mayors do not want to eliminate all revenue. Thus, pricing versus outcomes is a key trade-off.

Another key trade-off relates to the impacts of drivers' decisions on other modes of transportation. The pricing, as well as convenience, of public transportation will affect its use. Pricing levels and convenience of use will also affect the use of alternative routes. There is significant risk of just moving the congestion problems elsewhere. Thus, a key trade-off involves levels of incentives for alternative choices and the impacts of these incentives across all alternatives.

Step 5: Identify alternative representations of these phenomena. Three representations are needed. The route structure of the city can be represented as a directed graph, with directions of arcs indicating two-way and one-way streets. A traffic flow model is needed to predict the impacts of demands on flows along arcs, as well as the impacts of these flows on congestion. Third,

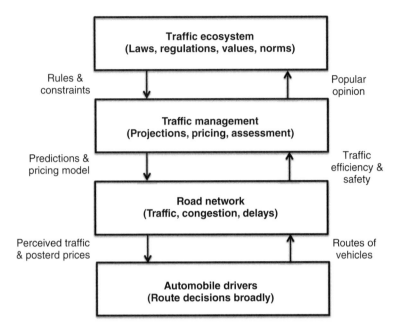

Figure 2.1   Hierarchical Visualization of Congestion Pricing Problem

an agent-based model of drivers' decision making is needed to model drivers' choices relative to pricing levels and alternative modes and routes of travel.

As shown in Figure 2.1, these phenomena can be viewed as occurring at four levels or scales – people, process, organization, and ecosystem. The people level includes drivers and their choices. The process level includes traffic flows on the city's route structure. The organization level includes the agencies managing the transportation infrastructure, and perhaps city government, central city businesses. The ecosystem level concerns the legal and social rules of the game.

The initial driver agent models might be simply probabilities of avoiding the central city as a function of prices. However, this representation will quickly be inadequate for two reasons. First, drivers' choices of alternatives will depend on what it is assumed they know. If we assume that they all have smart phones, they will know quite a bit. Consequently, the agent models will have to include mechanisms for using this information to make decisions.

The second shortcoming of the initial simple model is its lack of depiction of the social networks among drivers. Drivers are well known for quickly identifying and sharing shortcuts with friends and family. It should be expected that drivers would quickly discover holes in any pricing models and exploit these opportunities. Models that provide mechanisms for such discoveries will be invaluable for uncovering unintended consequences.

Step 6: Assess the ability to connect alternative representations. The route structure model will define what routes are feasible. The traffic flow model will predict the spatial flows (vehicles per unit time) for each arc in the network at each point in time, including congestion levels for each arc. The agent-based model will predict the decisions of each driver as a function of pricing and any other information they are assumed to have.

To connect these models, a mechanism is needed to translate thousands of individual decisions into consequent vehicle flows. More specifically, we need to connect a large number of agent-based models of drivers' decision making with partial differential equation models of vehicle flow and congestion. Thus, we need to somehow combine thousands of decisions about vehicle routes, as well as execution of those routes, with models that represent flow spatially in terms of vehicles per unit time.

This combination problem might be addressed heuristically were it not for the fact that drivers' perceptions of flows will influence their decisions. Thus, put simply, decisions affect flows, and flows affect decisions – all across thousands of agents. Thus, we need to translate back and forth between the two rather different types of representation. This will require careful thought – see Chapter 8 for a discussion of model composition.

Another connection issue concerns not only the movements of variables among the component representations of the overall model, but also the fact that perceptions of variables, correct or not, will influence agents' decisions. This might be handled by providing agent models with rudimentary perceptual capabilities – see Chapter 5 for a discussion of mental models. As a consequence, agents would be acting on approximate or perhaps even distorted information from the process level.

Step 7: Determine a consistent set of assumptions. The accuracy of the predicted levels of congestion will be highly affected by the validity of assumptions about agents' sources of information, perceptions of information, and decision-making rules. Making this even more complicated is the fact that drivers will differ along these dimensions. Thus, the accuracy of assumptions about agents will be affected by the accuracy of the distributional assumptions about these dimensions. This is an area where sensitivity analyses will be invaluable.

Step 8: Identify data sets to support parameterization. The topic of traffic congestion in general, and congestion pricing in particular, has been heavily studied. Thus, there are likely to be substantial data sets that can be accessed. However, the validity of these data to the particular city and driver population of interest may be questionable. Nevertheless, this is the right place to start, at least in terms of prototyping an initial model and exploring the parameter sensitivity of model predictions to key parameters.

Step 9, *Program and verify computational instantiations*, and step 10, *Validate model predictions, at least against baseline data*, were not performed for this example.

SUPPORTING THE METHODOLOGY

Support of the aforementioned methodology is discussed in detail in Chapters 8–10. At this point, however, it is useful to summarize the vision of how this support should function.

The support system is an interactive environment that enables pursuit of the set of nominal steps outlined earlier. These steps are "nominal" in that users are not required to follow them. Advice is provided in terms of explanations of each step and recommendations for methods and tools that might be of use. This advice can take several forms. At one extreme, it would be copies of this book, as well as referenced publications. At the other extreme, expert advice and automation might involve streamlined execution of the steps of the methodology. I think a middle ground is more acceptable to users while also being technically feasible.

Compilations of physical, human, economic, and social phenomena are available, including standard representations of these phenomena, in terms of equations, curves, surfaces, and son, but not software. Advice is provided in terms of variable definitions, units of measure, and so on, as well typical approximations, corrections, and so on. Of particular importance, advice is provided on how to meaningfully connect different representations of phenomena.

Visualization tools are available, including block diagrams, IDEF (Integration DEFinition), influence diagrams, and systemigrams (see Chapter 8). Software tools for computational representations are recommended, with emphasis on commercial off-the-shelf platforms that allow input from and export to, for example, Microsoft Excel and Matlab. Examples include AnyLogic, NetLogo, Repast, Simio, Stella, and Vensim (see Chapter 9).

In keeping with contemporary approaches to user support, this methodological support is not embodied in a monolithic software application. Instead, this support framework operates as a fairly slim application that assumes that users have access to rich and varied toolsets elsewhere on their desktops. The support provides structured guidance on how to best use this toolset to address the phenomena associated with the problem that has motivated the modeling effort at hand.

We assume that model development occurs within the confines of one or more desktops or laptops. We also envision capabilities to export interactive visualizations to highly immersive simulation settings as discussed in Chapter 8. Methodological support will not, however, be premised on users having access to such settings.

CONCLUSIONS

This chapter began with a discussion of human-centered methods and tools and the issues of viability, acceptability, and validity. We next discussed six problems archetypes that are addressed throughout this book. These problems provide the use cases for the methodology. Attention then turned to the overall 10-step methodology. An example use of the methodology was discussed. The chapter concluded with consideration of how use of the methodology can be supported.

REFERENCES

Pennock, M.J., & Rouse, W.B. (2014a). The challenges of modeling enterprise systems. *Proceedings of the 4ᵗʰ International Symposium on Engineering Systems.* Hoboken, NJ, June 8–11.

Pennock, M.J., & Rouse, W.B. (2014b). Why connecting theories together may not work: How to address complex paradigm-spanning questions. *Proceedings of the 2014 IEEE International Conference on Systems, Man and Cybernetics,* San Diego, Oct 5–8.

Rouse, W.B. (1991). *Design for Success: A Human-Centered Approach to Designing Successful Products and Systems.* New York: Wiley.

Rouse, W.B. (1998). Computer support of collaborative planning. *Journal of the American Society for Information Science,* 49 (9), 832–839.

Rouse, W.B. (2007). *People and Organizations: Explorations of Human-Centered Design.* New York: Wiley.

Rouse, W.B. (2014). Human interaction with policy flight simulators. *Journal of Applied Ergonomics,* 45 (1), 72–77.

3

PERSPECTIVES ON PHENOMENA

INTRODUCTION

The notion of phenomena is central to the approach to understanding complex systems and enterprises advanced in this book. Identification, composition, and visualization of phenomena are central to the first four steps on the methodology presented in Chapter 2. Representation of phenomena, composition of these representations, and computation and visualization of the resulting compositions are central to the last six steps of the methodology presented in Chapter 2. In this chapter, we explore the fundamental nature of phenomena, both historically and from contemporary perspectives. A taxonomy of phenomena is introduced. Application of the taxonomy is illustrated for categorizing phenomena associated with the six archetypal problems introduced in Chapter 2. Finally, visualization of phenomena is discussed.

DEFINITIONS

First, what is meant by the term "phenomena"? There are many definitions that depend on the context of use – for example, a particular jazz

Modeling and Visualization of Complex Systems and Enterprises:
Explorations of Physical, Human, Economic, and Social Phenomena, First Edition. William B. Rouse.
© 2015 John Wiley & Sons, Inc. Published 2015 by John Wiley & Sons, Inc.

saxophonist might be characterized as a "phenomenon." Within the context of science and engineering, Boon (2012) provides examples such as "elasticity, specific weight, viscosity, specific heat content, melting point, electrical resistance, thermal conductivity, magnetic permeability, physical hysteresis, crystallinity, refractivity, chemical affinity, wavelength, chemical diffusivity, solubility, electric field strength, super conductivity, and atomic force." These examples fit into the framework elaborated later in this chapter, but form only a small part of the spectrum of phenomena of interest in this book.

Boon (2012) explores the role that phenomena play in science and engineering. Citing Hacking (1983) and Bogen and Woodward (1988), she indicates that phenomena motivate scientific theories and serve as evidence for theories. In contrast, in engineering science, phenomena are important in that they can play a productive or obstructive role in the technologically functioning of artifacts. She argues that, "Scientific concepts of phenomena function as epistemic tools for creating and intervening with phenomena that are of technological relevance." In other words, "We purposefully aim at creating phenomena, not only because of their epistemic role, but also because we want these phenomena for specific technological functions."

Discussing the nature of engineering, she notes that, "Conceptualization (of) phenomena that we wish to create or intervene with for performing technological functions involves fitting together relevant but heterogenous content." Such "fitting together," she characterizes as design. Pursuing the notion of design, she concludes, "The ultimate challenge of designing is not a correct representation of the designed object, but an adequate fit of heterogeneous epistemic content concerning real world aspects of the object, such as its structure, the materials used, the construction techniques needed, the measures taken for its safety, robustness, etc."

Her colleague, Van Ommerren (2011), reviews a wide range of perspectives on and definitions of phenomena. He illustrates his points with case studies that deal with heat transfer in fiber-reinforced composites. He concludes that "A phenomenon in engineering science is something that is used in service of the target system. A phenomenon is an instrument which can provide knowledge about a target system and which can be used to intervene with a target system."

Arthur (2009) explores the meaning of the term technology within the overall concept of technological innovation. He defines technology as a collection of phenomena (physical, behavioral, or organizational) captured and put to use, that is, programmed to our purposes. Such "purposed systems" can include aircraft, symphonies, and contracts.

Arthur argues that science seeks to discover and understand the nature of phenomena, particularly at a scale or in a world not open to direct human observation. Technologists use science because it is the only way to understand how phenomena work at deeper layers. Science is the probing of phenomena via technology.

More broadly, he defines a domain as "any cluster of components drawn from in order to form devices or methods, along with its collection of practices and knowledge, its rules of combination, and its associated ways of thinking." He asserts that "Half of the effectiveness of a domain lies in its reach – the possibilities it opens up; the other half lies in using similar combinations again and again for different purposes." Most broadly, he defines an economy as the set of arrangements and activities by which a society satisfies its needs.

As an illustration of these concepts, Arthur discusses "the enabling technologies of electrification, the electric motor and generator, which arrived in the 1870s, but the full effects on industry were not felt until well into the 1910s and 1920s." Electric motors became available in the 1880s, but it took factories four decades to replace steam power with electric power because factories had to be re-architected.

More broadly, inventors learned how to harness, control, and design phenomena ranging from steam to electricity, and from internal combustion to aerodynamic lift to analog and digital computation. Innovators found ways to bring these controlled phenomena to the marketplace to provide value to people. This process happened and continues to happen, again and again (Rouse, 1992, 1996, 2014). According to Mokyr (1992), this process of technological innovation provides society with a "lever of riches."

How does technology transfer to market innovation? This has been a question of great interest for several decades. Burke (1996) provides an historical perspective on this question. He chronicles the evolution of roughly 300 technologies over a period of 500 years. He reaches two all-embracing conclusions. First, the application envisioned for new technology is hardly ever the application on which the technology has its greatest impact. Second, the original investor in a new technology is hardly ever the recipient of the greatest returns enabled by the technology. Interest in the management of innovation has been motivated by desires to avoid this destiny.

How do enterprises deal with fundamental changes in their markets, technologies, and so on (Rouse, 1996, 2006)? Most enterprises, in fact almost all enterprises, have great difficulty fundamentally changing their business models and their technology paradigms. Creative destruction (Schumpter, 1942) is typically the result. Companies refine their business models and technologies until they dominate their market. Then, change happens and they become irrelevant and disappear. This is great for the economy, but a bit rough on companies.

Disruptions that lead to creative destruction need not destroy one market and create another (Christensen, 1997). Ships, trains, automobiles, and airplanes look pretty much now as they did decades ago. They are more reliable and efficient, as well as laced with electronics and computers. However, these improvements do not constitute innovations in the sense that the markets involved are not changing what they buy. New competitors may be winning, but they are just selling better and cheaper versions of what the markets were already buying. This is certainly disruptive, but not necessarily innovative.

Succinctly, technology involves harnessing phenomena to purposeful ends. Often, phenomena are first characterized and then understood by science. Engineering, and inventors in particular, determine how to harness the phenomena, often long before scientists have determined how best to characterize the phenomena. Innovation results when the value proposition for delivering the benefits of the phenomena makes sense in a marketplace. Eventually, this value proposition is disrupted, new technologies displace the old, and creative destruction moves the economy and society forward. With this background, we can now focus squarely on the nature of phenomena.

HISTORICAL PERSPECTIVES

The history of harnessing phenomena to purposeful ends is rich with examples of ingenuity and determination. In some cases, it took hundreds and even thousands of years from recognition and understanding of a phenomenon to fully leveraging it for useful ends.

Steam to Steamboats

Greek philosophers argued that the four fundamental natural phenomena were earth, air, fire, and water. Much was accomplished with these phenomena. Water and fire were combined to yield steam. By the 1st century AD, people had discovered that water heated in a closed vessel produced high-pressure steam that could be used to power various novelty devices. Channeling the pressure using a cylinder led to Thomas Newcomen's steam engine in 1712. James Watt added an external condenser in 1776 so that condensation did not result in significant losses of heat. Robert Fulton used Watt's engine and a hull designed for this mode of propulsion to develop the first commercially successful steamboat, the Clermont, in 1807. Thus, steam to steamboats took roughly 1800 years.

Wind to Wings

Air in the form of wind is a phenomenon that was exploited quite early. Wind plus cloth sails led to sailboats by 4000 BC. Wind plus vanes (frameworks for sails) plus wooden gears led to windmills by 1000 AD. Wind plus lift plus thrust led to heavier than air flight in 1800–1900. Thrust is needed to create movement, and hence wind, to enable the lifting phenomena of airfoils. Thus, powered flight emerged in 1903. Wind to wings took almost 5000 years.

Electricity to Electric Lights

Electricity was not one of the original four fundamental phenomena. However, static electricity was known by 600 BC. Magnetism was known then as well, but it was not until 1600 that William Gilbert outlined the relationship between electricity and magnetism, which James Clerk Maxwell formalized in 1873. Earlier, in 1752, Benjamin Franklin combined lightning, a wetted kite string, and a key to demonstrate the phenomenon of electricity. By 1802, Humphrey Davy had shown that electricity when passed through platinum produced light. Warren de la Rue enclosed a coiled platinum filament in a vacuum tube and passed an electric current through it in 1840 for the first light bulb.

In 1878, Thomas Edison created a light bulb with a more effective incandescent material, a better vacuum, and high resistance that made power distribution from a centralized source economically viable. His Pearl Street Station in New York City meant that houses in that area only had to install electric lights. They did not have to produce their own electricity. This made adoption of his innovation much easier. The path from knowledge of electricity to viable electric lighting systems took almost 2500 years. In all three of these examples, harnessing known phenomena to useful ends took many centuries.

Macro and Micro Physics

Other phenomena of historical interest include the inventions of Leonardo da Vinci (1452–1519), harnessing motion, force, and flow for his civil and military work. Isaac Newton (1643–1727) studied the motions of planets, and created calculus in the process, although Gottfried Leibniz (1643–1716) claimed credit for calculus as well. Albert Einstein (1879–1955) articulated a special theory of relativity in 1905 and a general theory in 1916, thereby relegating Newton's theory of mechanics to the status of a special case for objects whose movements are much slower than the speed of light. Special relativity concerns interactions of elementary particles while general relativity applies to phenomena of cosmology and astrophysics.

Behaviors of elementary particles became phenomena of much scientific interest as the nineteenth century gave way to the twentieth. Joseph Thomson (1856–1940) discovered the electron in 1897. Ernest Rutherford (1871–1937) discovered the nucleus in 1909. Niels Bohr (1885–1962) added the notion of fixed orbits of the electron in 1913. Louis de Broglie (1892–1987) contributed the wave theory of subatomic particles in 1924, which was extended by Erwin Schrödinger (1887–1961) in 1926. The inherent uncertainties associated with wave behaviors led Werner Heisenberg (1901–1976) to articulate his uncertainty principle in 1927.

Probability and Utility

Uncertainty had long fascinated philosophers, mathematicians, and physicists. Probability theory developed as a means to address uncertainties surrounding many phenomena. Mathematical methods associated with the notion of probability arose in the correspondence of Pierre de Fermat (1601–1666) and Blaise Pascal (1623–1662). For example, they developed the construct of the expected value in 1654, although they never published their findings. Christiaan Huygens (1629–1695) published a similar result in 1657.

Jeremy Bentham (1748–1832), Thomas Bayes (1702–1761), and Daniel Bernoulli (1700–1782) introduced the concept of "utility" to explain the nature of human economic decision making. In 1738, Bernoulli articulated the theory that expected utility, that is, the expected value of the utility function, should govern choices. It was 1947 before John von Neumann (1903–1957) and Oskar Morgenstern (1902–1977) provided a mathematical proof of this theory based on a core set of axioms of rationality.

CONTEMPORARY PERSPECTIVES

In this section, we consider contemporary views of phenomena from the perspectives of both science and engineering. Science focuses on decomposition of phenomena into increasingly fine-grained representations of reality. In contrast, engineering addresses the composition of representations of myriad phenomena into integrated solutions of difficult large-scale problems.

Four Fundamental Forces

Thoughts about fundamental phenomena have come a long way since earth, air, fire, and water had this honor. Contemporary physicists and other scientists see fundamental natural as involving four constructs – gravitation, electromagnetism, strong nuclear force, and weak nuclear force. Gravitation and electromagnetism are the only two we experience in everyday life. The strong and weak nuclear forces operate at distances of femtometers (10^{-15} m) or less.

Gravitation affects macroscopic objects over macroscopic distances. It is the only force that acts on all particles having mass and has an infinite range. Gravity cannot be absorbed, transformed, or shielded against. Gravity always attracts; it never repels.

Electromagnetism is the force that acts between electrically charged particles. This phenomenon includes the electrostatic force acting between charged particles at rest, as well as the effects of electric and magnetic forces acting between charged particles as they move relative to each other.

The strong and weak nuclear forces pertain to phenomena within the atomic nucleus. The strong force varies with distance and is practically unobservable at distances greater than ten 10^{-15} m. The weak nuclear force is responsible for some nuclear phenomena such as beta decay.

This notion that these four fundamental phenomena underlie everything does not seem as compelling as earth, air, fire, and water, as least from an engineering perspective. Yet, understanding, predicting, and then controlling these forces may, in some distant future, lead to amazing inventions and subsequent innovations. This will likely require a level of maturity that engineering has achieved in several contemporary areas.

Computational Fluid Dynamics

Contemporary perspectives on the engineering of phenomena to purposeful ends are best illustrated by considering three design problems where approaches have been codified and virtually standardized. Computational fluid dynamics deals with the mechanics of fluids, for example, water and air, as they interact with the built environment such as buildings or aircraft. Computational fluid dynamics employs numerical methods and algorithms to solve and analyze problems that involve such fluid flows. Computational methods are used to simulate the interaction of liquids and gases with surfaces defined by boundary conditions representing the built environment. These methods substantially reduce, but do not eliminate, needs for empirical studies in towing tanks or wind tunnels. Chapter 4 discusses a good example of this approach, namely, urban oceanography.

Integrated Circuit Design

Another "standard" engineering problem is integrated circuit design, which involves formulation of a plan for fabricating electronic components, such as transistors, resistors, and capacitors, as well as the metallic interconnections of these components, onto a piece of semiconductor, usually silicon. The individual components must be isolated as the substrate silicon is conductive and often forms an active region of the individual components. Power dissipation of transistors and interconnect resistances is also a significant design issue. The physical layout of circuits is a central design issue. Layout affects speed of operation, segregation of noisy portions of the chip from quiet portions, balancing effects of heat generation across the chip, and facilitating the placement of connections to circuitry outside the chip. There are 20 or so standard software tools available to support these design tasks, with a wide range of prices, and some open source and free.

Supply Chain Management

A third "standard" engineering problem is supply chain management. This refers to the set of activities associated with executing supply chain transactions, managing supplier relationships, and controlling related business processes. More specifically, this includes customer order processing, purchase order processing, inventory management, goods receipt, warehouse management, and supplier sourcing and management, as well as forecasting the attributes of these activities. Balancing the disparity between supply and demand is accomplished by the design of business processes and use of optimization algorithms to allocate and manage resources. Supply chain management software usually enables organizations to trade electronically with supply chain partners. There is a large number of software tools and consulting companies available to support any or all of the aforementioned supply chain management activities.

Summary

These three examples of standard engineering problems are at the other end of the scale from science's four fundamental forces. They address the composition of representations of myriad phenomena into integrated solutions of difficult large-scale problems. In contrast, science focuses on decomposition of phenomena into increasingly fine-grained representations of reality. Thus, the scale that differentiates science and engineering runs from narrow decomposition of phenomena to broad composition of phenomena.

TAXONOMY OF PHENOMENA

We need a taxonomy of phenomena in order to address the six archetypal problems introduced in Chapter 2. This taxonomy needs to include

TABLE 3.1   Class of Phenomena versus Example Phenomena of Interest

Class of Phenomena	Example Phenomena of Interest
Physical, natural	Temporal and spatial relationships and responses
Physical, designed	Input–output relationships, responses, stability
Human, individuals	Task behaviors and performance, mental models
Human, teams and groups	Team and group behavior and performance
Economic, micro	Consumer value, pricing, production economics
Economic, macro	Gross production, employment, inflation, taxation
Social, organizational	Structures, roles, information, resources
Social, societal	Castes, constituencies, coalitions, negotiations

phenomena at the various levels of abstraction and aggregation associated with these problems. The taxonomy also serves as the organizing model for Chapters 4–7.

Table 3.1 summarizes the eight classes of phenomena addressed in this book.

A few caveats are in order. First, this taxonomy is not a general taxonomy of all possible phenomena. For example, physicists and chemists would find many of their interests unrepresented. Put simply, the nature of this taxonomy was determined by the phenomena prevalent in the six archetypal problems, examples of which appear in the right column of Table 3.1.

Second, this taxonomy indicates human phenomena[1] as representing two of eight classes of phenomena. Viewed more broadly, however, human phenomena are laced throughout at least seven of the classes and, if human biology is considered, all eight classes. Biological, physiological, psychological, economic, sociological, and anthropological phenomena are pervasive across the archetypal problems discussed in this book. Yet, we need an organizing scheme to keep the exposition digestible.

Figure 3.1 suggests one way of looking at the relationships among the classes in the taxonomy. This depiction shows four broader classes, each of which involves humans in many ways. There are, of course, other ways that these four classes could connect. Fortunately, the organization of the content in Chapters 4–7 does not dictate how the phenomena discussed are represented and composed into overall visualizations or computational models in Chapters 8–10.

---

[1] An alternative labeling of this class might be "behavior," but this is at least as problematic as "human" because the phenomena in all the other classes also exhibit behaviors.

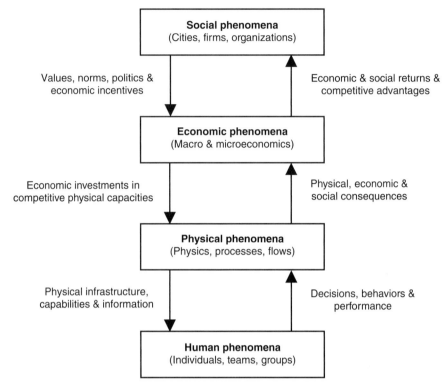

Figure 3.1   Hierarchy of Phenomena

Behavioral and Social Systems

Harvey and Reed (1997) provide a valuable construct for a broad perspective on behavioral and social systems. They provide a compelling analysis of levels of abstraction of social systems versus viable methodological approaches. Table 3.2 provides a version of this construct, modified to better support the line of reasoning in this book. The appropriate matches of levels of abstraction and viable approaches are highlighted in black.

At one extreme, the evolution of social systems over long periods of time is best addressed using historical narratives. It does not make sense to aspire to explain history, at any meaningful depth, with a set of equations. At the other extreme, it makes great sense to explain physical regularities of the universe using predictive and statistical modeling.

Considering the hierarchy of phenomena in Figure 3.1, the types of problems addressed in this book often require addressing physical, human,

TABLE 3.2   Hierarchy of Complexity versus Approaches (Adapted from Harvey & Reed, 1997)

Hierarchy of Complexity in Social Systems	Predictive Modeling	Statistical Modeling	Iconological Modeling	Structural Modeling	Ideal Type Modeling	Historical Narratives
Evolution of social systems over decades, centuries, and so on.						■
Competition and conflict among social systems						■
Cultural dominance and subcultural bases of resistance					■	■
Evolution of dominant and contrary points of view					■	■
Interorganizational allocations of power and resources				■	■	■
Personal conformity and commitment to roles and norms			■	■	■	■
Intraorganizational allocation of roles and resources		■	■	■	■	■
Distribution of material rewards and esteem		■	■	■	■	
Division of labor in productive activities		■	■	■	■	

(*continued*)

TABLE 3.2 *(Continued)*

Hierarchy of Complexity in Social Systems	Predictive Modeling	Statistical Modeling	Iconological Modeling	Structural Modeling	Ideal Type Modeling	Historical Narratives
Social infrastructure of organizations	■	■	■	■	■	
Ecological emergence of designed organizations	■	■	■	■	■	
Ecological organization of living phenotypes	■	■	■			
Evolution of living phenotypes	■	■	■			
Regularities of physical universe	■	■				

economic, and social phenomena. Consequently, we need to operate at more than one level of Table 3.2. This implies that our overall representation of the problem – either visually or computationally – will involve multiple types of representations. This poses many challenges that are addressed in Chapters 8–10.

Problems versus Phenomena

Table 3.3 summarizes the phenomena associated with the six archetypal problems introduced in Chapter 2 and discussed in Chapters 4–7. The eight classes of Table 3.1 include the central phenomena that need to be addressed in order to address the archetypal problems. This, of course, is not surprising as the problem description drove the design of the taxonomy. Further, the six problems were designed to include phenomena far broader than just physical phenomena.

VISUALIZING PHENOMENA

Step 3 of the methodology introduced in Chapter 2 focuses on visually representing phenomena and relationships among phenomena. Step 4 employs

TABLE 3.3   Phenomena Associated with Archetypal Problems

Problem	Phenomena	Chapter	Category
Counterfeit parts	Physical flow of parts supply, including inspection	4	Physical, designed
Counterfeit parts	Physical assembly of parts and system performance	4	Physical, designed
Healthcare delivery	Physical flow of demands to capacities, yielding revenues and health outcomes	4	Physical, designed
Traffic control	Physical traffic congestion throughout urban transportation grid	4	Physical, designed
Traffic control	Physical flow of vehicles through urban transportation grid	4	Physical, designed
Urban resilience	Physical responses of urban infrastructures	4	Physical, designed
Cancer biology	Intercellular biological signaling mechanisms throughout body	4	Physical, natural
Cancer biology	Intracellular biological signaling mechanisms within cells	4	Physical, natural
Cancer biology	Mechanisms of cell growth and death, driven by genetics and behaviors	4	Physical, natural
Healthcare delivery	Disease incidence and progression of stratified population	4	Physical, natural
Urban resilience	Physical flow of water through urban topography	4	Physical, natural
Cancer biology	Human decisions on diagnoses given test results	5	Human, individual
Cancer biology	Human decisions to employ drugs targeted at signaling mechanisms	5	Human, individual

(continued)

TABLE 3.3 (*Continued*)

Problem	Phenomena	Chapter	Category
Financial system	Human decisions by investors relative to risks and rewards	5	Human, individual
Traffic control	Human decisions by individual drivers on where and when to drive	5	Human, individual
Traffic control	Human performance in manual control of vehicle	5	Human, individual
Urban resilience	Human perceptions, expectations, and intentions	5	Human, individual
Counterfeit parts	Human decisions by supplier executive teams	5	Human, team or group
Healthcare delivery	Human decisions of clinical teams on use of capacities	5	Human, team or group
Traffic control	Human decisions by traffic management teams on pricing	5	Human, team or group
Urban resilience	Human decisions of urban teams on communications	5	Human, team or group
Counterfeit parts	Macroeconomics of defense acquisition	6	Economic, macroeconomics
Financial system	Macroeconomics of demand and supply of investment products in general	6	Economic, macroeconomics
Healthcare delivery	Macroeconomics of demand, supply, and payment practices for healthcare	6	Economic, macroeconomics
Counterfeit parts	Microeconomics of supplier firms, including testing	6	Economic, microeconomics
Financial system	Microeconomics of firms creating specific investment products	6	Economic, microeconomics

TABLE 3.3 (*Continued*)

Problem	Phenomena	Chapter	Category
Financial system	Microeconomics of risk and return characteristics of specific investment products	6	Economic, microeconomics
Financial system	Microeconomics of investor demand and prices of specific investment products	6	Economic, microeconomics
Healthcare delivery	Microeconomics of providers' investments in capacities to serve demand	6	Economic, microeconomics
Traffic control	Drivers information sharing on routes and shortcuts	7	Social, information sharing
Urban resilience	Peoples information sharing on perceptions, expectations, and intentions	7	Social, information sharing
Counterfeit parts	Social system of defense industry	7	Social, organizations
Healthcare delivery	Social system of healthcare industry	7	Social, organizations
Urban resilience	Social system of city, communities, and neighborhoods	7	Social, organizations
Counterfeit parts	Values and norms of defense industry	7	Social, values and norms
Healthcare delivery	Values and norms of healthcare industry	7	Social, values and norms
Urban resilience	Values and norms of communities and neighborhoods	7	Social, values and norms

these visualizations to identify key trade-offs, which may or may not warrant deeper computational explorations. Chapters 8 and 9 consider how initial visualizations can be translated into equations and computational representations.

It is essential to the methodology advocated in this book that one not jump immediately from the question of interest to some form of mathematics. As discussed in Chapter 9, each of the alternative mathematical formalisms carries with it a set of standard assumptions (e.g., linearity, independence, continuity) that may be lost in the translation process. For this reason and others, it is important to begin with more basic visualizations.

Chapter 4 on physical phenomena includes examples on human biology, urban oceanography, vehicle powertrains, manufacturing processes, and counterfeit parts. These examples are visualized using block diagrams or flow charts. The biological example employs a hierarchical block diagram.

Chapter 5 on human phenomena considers examples of manual control, problem solving, multitask decision making, traffic control via congestion pricing, mental models of individual and teams, and performing arts teams. These examples are visualized using block diagrams, flow charts, and iconographs.

Chapter 6 on economic phenomena addresses examples of production economics, consumer economics, gross domestic product, and healthcare delivery. Visualizations include flowcharts and hierarchical block diagrams, with a few classic equations where the concepts are inherently numeric.

Chapter 7 on social phenomena considers examples of earth as a system, castes and outcastes, acquisition as a game, port and airport evacuation, emergence of cities, and urban resilience. Visualizations include block diagrams, flowcharts, and network diagrams, with a couple of classic equations.

All of these initial visualizations have a few characteristics in common. They portray phenomena central to the question being addressed, as well as relationships among these phenomena. Many of these relationships are of the simple form "X affects Y." Some of the relationships have more semantic meaning – for example, "X feeds, controls, or constrains Y." Of most importance, these visualizations are immediately understandable.

CONCLUSIONS

In this chapter, we explored the fundamental nature of phenomena, both historically and from contemporary perspectives. A taxonomy of phenomena was introduced. Use of the taxonomy was illustrated using the six archetypal problems introduced in Chapter 2. Finally, visualization of phenomena was discussed. Chapters 4–7 consider physical, human, economic, and social phenomena, respectively. As noted earlier, these chapters are laced with a wealth of examples. In Chapters 8–10, we return to the use of this knowledge of phenomena in the overall methodology, as well as discussions of understanding and supporting human problem solving.

REFERENCES

Arthur, W.B. (2009). *The Nature of Technology: What it is and How it Evolves*. New York: Free Press.

Bogen, J., & Woodward, J. (1988). Saving the phenomena. *Philosophical Review*, 97 (2), 303–352.

Boon, M. (2012). *Scientific Concepts in the Engineering Sciences: Epistemic Tools for Creating and Intervening with Phenomena*. Enschede, The Netherlands: University of Twente, Department of Philosophy.

Burke, J. (1996). *The Pinball Effect: How Renaissance Water Gardens Made the Carburetor Possible and Other Journeys Through Knowledge*. Boston: Little, Brown.

Christensen, C.M. (1997). *The Innovator's Dilemma: When New Technologies Cause Great Firms to Fail*. Boston: Harvard Business School Press.

Hacking, I (1983). *Representing and Intervening: Introductory Topics in the Philosophy of Natural Science*. Cambridge, UK: Cambridge University Press.

Harvey, D.L., & Reed, M. (1997). Social science as the study of complex systems. In L. D. Kiel & E. Elliot, Eds., *Chaos Theory in the Social Sciences: Foundations and Applications (Chap. 13)*. Ann Arbor: The University of Michigan Press.

Mokyr, J. (1992). *The Lever of Riches: Technological Creativity and Economic Progress*. New York: Oxford University Press.

Rouse, W.B. (1992). *Strategies for Innovation: Creating Successful Products, Systems and Organizations*. New York: Wiley.

Rouse, W.B. (1996). *Start Where You Are: Matching Your Strategy to Your Marketplace*. San Francisco: Jossey-Bass.

Rouse, W.B. (Ed.). (2006). *Enterprise Transformation: Understanding and Enabling Fundamental Change*. New York: Wilcy.

Rouse, W.B. (2014). *A Century of Innovation: From Wooden Sailing Ships to Electric Railways, Computers, Space Travel and Internet*. Raleigh, NC: Lulu Press.

Schumpter, J. (1942). *Capitalism, Socialism, and Democracy*. New York: Harper

Van Ommerren, E.J. (2011). *Phenomena in Engineering Science*. Enschede, The Netherlands: University of Twente, Department of Philosophy.

4

PHYSICAL PHENOMENA

INTRODUCTION

Table 4.1 lists the physical phenomena of interest for the six archetypal problems introduced in Chapter 2 and discussed in Chapter 3. All of these phenomena are concerned with physical flow, responses, signaling, and control. This chapter addresses the state of knowledge in representing these phenomena.

This chapter considers two types of physical phenomena. First, we discuss naturally occurring phenomena such as biological signaling, weather, and water flow. We consider two examples – human biology and urban oceanography. Then we address designed phenomena such as systems engineered to move people and goods. Examples of interest here include vehicle powertrains and manufacturing processes. Of course, we are often concerned with the intersection of designed and natural physical phenomena. The chapter concludes with an elaboration of the counterfeit parts problem.

NATURAL PHENOMENA

Physical phenomena associated with the natural world include weather, water and air flow, ice formation and melting, climate change in general, and many

Modeling and Visualization of Complex Systems and Enterprises:
Explorations of Physical, Human, Economic, and Social Phenomena, First Edition. William B. Rouse.
© 2015 John Wiley & Sons, Inc. Published 2015 by John Wiley & Sons, Inc.

TABLE 4.1   Physical Phenomena Associated with Archetypal Problems

Problem	Phenomena	Category
Counterfeit parts	Physical flow of parts supply, including inspection	Physical, designed
Counterfeit parts	Physical assembly of parts and system performance	Physical, designed
Healthcare delivery	Physical flow of demands to capacities, yielding revenues and health outcomes	Physical, designed
Traffic control	Physical traffic congestion throughout urban transportation grid	Physical, designed
Traffic control	Physical flow of vehicles through urban transportation grid	Physical, designed
Urban resilience	Physical responses of urban infrastructures	Physical, designed
Cancer biology	Intercellular biological signaling mechanisms throughout body	Physical, natural
Cancer biology	Intracellular biological signaling mechanisms within cells	Physical, natural
Cancer biology	Mechanisms of cell growth and death, driven by genetics and behaviors	Physical, natural
Healthcare delivery	Disease incidence and progression of stratified population	Physical, natural
Urban resilience	Physical flow of water through urban topography	Physical, natural

others. Models of physical phenomena are typically concerned with transformation and flow of matter and energy over space and time. Common representations of such phenomena include partial differential equations. For example, Maxwell's equations are a set of partial differential equations that underlie electrodynamics, optics, and electric circuits. The Navier–Stokes equations are a set of partial differential equations that describe the motion of fluid substances. Laplace's equation, the diffusion equation, and the wave equation are

TABLE 4.2   Phenomena Associated with the Natural Sciences

Discipline	Phenomena Addressed
Astronomy	Celestial bodies and their interactions in space
Atmospheric science	Atmosphere, its processes, the effects of other systems on the atmosphere, and the effects of the atmosphere on these other systems
Biology	Phenomena related to living organisms, at scales of study ranging from subcomponent biophysics to complex ecologies
Chemistry	Phenomena associated with the structure, composition, and energetics of matter as well as the changes it undergoes
Earth science	Atmosphere, hydrosphere, oceans, and biosphere, as well as the solid earth
Physical geography	Natural environment and how the climate, vegetation and life, soil, oceans, water, and landforms are produced and interact
Physics	Physical laws of matter, energy, and the fundamental forces of nature that govern the interactions between particles and physical entities (such as planets, molecules, atoms, or the subatomic particles)

other instances applicable to phenomena ranging from heat transfer to sound propagation.

Physical phenomena are addressed as ends in themselves by the natural sciences. These sciences are summarized in Table 4.2. Physics is the classic science of Galileo Galilei, Isaac Newton, Albert Einstein, and Stephen Hawking although they can also be claimed by astronomy for their studies of celestial bodies and their interactions in space. Antoine Lavoisier, Joseph Priestly, John Dalton, and Marie Curie are among the great chemists. Charles Darwin, Gregor Mendel, Louis Pasteur, and Barbara McClintock are among the most well-known biologists. Lavoisier and Priestly are often claimed by biology as well for their research on oxygen.

Thus, we see that many of the greatest scientists do not fit neatly into one disciplinary category. They were often interested in phenomena that span current disciplines. Indeed, the intense specialization typical of current science has only been the hallmark of the natural sciences in recent decades, perhaps since World War II. For complex systems, phenomena that cut across academic disciplines are quite common. The physical world is not organized by academic disciplines. Such organization is convenient for the research community, but not an underlying distinction of physical reality.

Example – Human Biology

Hierarchical or multilevel models of biological phenomena are common, for example, (Ingber, 2003; Pavé, 2006). Figure 4.1 portrays human biology on six levels. Genes express proteins that result in cell growth. Cells make up tissues, which compose organs and serve as elements of various systems, all of which come together to form a specific human.

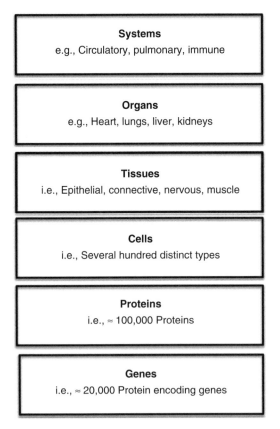

Figure 4.1   Multilevel Representation of Human Biological Phenomena

Genes express proteins that enable cellular phenomena, such as physical growth of cell structure (Ingber, 2003). These phenomena are, to an extent, controlled by signals both within and across cells. Cell signaling governs basic cellular activities and coordinates cell actions. The ability of cells to perceive and correctly respond to their microenvironment is the basis of development, tissue repair, and immunity as well as normal tissue homeostasis.

Most cell signals are chemical in nature. Growth factors, hormones, neurotransmitters, and extracellular matrix components are some of the many types of chemical signals cells use. These substances can exert their effects locally, or they might travel over long distances.

Once a receptor protein receives a signal, a series of biochemical reactions within the cell are launched. These intracellular signaling pathways typically amplify the message, producing multiple intracellular signals for every receptor. Activation of receptors can trigger the synthesis of small molecules called second messengers, which initiate and coordinate intracellular signaling pathways.

At any one time, a cell is receiving and responding to numerous signals, and multiple signal transduction pathways are operating. Many points of intersection exist among these pathways. Through this network of signaling pathways, the cell is constantly integrating all the information it receives from its external environment.

A key distinction for the multilevel representation in Figure 4.1 is that the lower three levels are intracellular, while the higher three levels involve both intra- and intercellular signaling. Autocrine signaling involves the cell directly affecting its own function by secreting substances that act on the cellular receptors. Such signaling is replicated across the ≈ 50 trillion cells in the human body.

The intercellular signaling includes paracrine signaling whereby cells can communicate with cells in the immediate environment, which is important in local immune response. The endocrine system refers to a collection of glands (e.g., pineal, pituitary, and thyroid) that secrete hormones directly into the circulatory system to signal distant organs.

Signaling and control is also enabled by the nervous system, a network of specialized cells that coordinate the actions of humans by sending signals from one part of its body to another. The central nervous system contains the brain and spinal cord. The peripheral nervous system consists of sensory neurons, groups of neurons termed ganglia, and nerves connecting them to each other and to the central nervous system.

The somatic nervous system controls voluntary muscular systems within the body. The autonomic nervous system, part of the peripheral nervous system, is a control system below the level of consciousness that controls such involuntary functions as heart rate and respiratory rate.

Sensory neurons are activated by inputs impinging on them from outside or inside the body, and send signals that inform the central nervous system of these inputs. Motor neurons, either in the central nervous system or in peripheral ganglia, connect neurons to muscles or other effector organs. Neural circuits are formed by the interaction of the different neurons. These circuits regulate humans' perceptions of the world as well as their bodies and behaviors.

The human body is constantly growing new cells while old cells die. In the process, almost most all cells are replaced within roughly 10 years. At the same time that this is happening, humans are walking, talking, learning, deciding, problem solving, acquiring assets, and socializing, behaviors that are discussed in later chapters. All in all, human biology is rather amazing.

Nevertheless, things do go wrong. One of these things is cancer (Mukherjee, 2010). Cancer emerges and reaches critical mass when there are signaling aberrations that fail to control misbehaving cells. Errors in cellular information processing are responsible for diseases such as cancer, autoimmunity, and diabetes.

Milella and her colleagues (2010) indicate that "Intercellular or intracellular signals are aberrantly sent and/or received, resulting in the uncontrolled proliferation, survival, and invasiveness of the cancer cell. The genomes of incipient cancer cells acquire mutant alleles of proto-oncogenes, tumor suppression genes, and other genes that control, directly or indirectly, cell proliferation, survival, and differentiation."

Citing Hahn and Weinberg (2002), they continue, "The pathogenesis of human cancers is governed by a set of genetic and biochemical rules that apply to most, and perhaps all, types of human tumors. These rules, in turn, reflect the operations of a few key intracellular regulatory circuits that operate on the majority of human cell types."

This understanding had led to the idea of using signaling pathways as therapeutic targets for drugs. Bianco and colleagues (2006) outline the phenomena being targeted, "Growth factor signals are propagated from the cell surface, through the action of trans-membrane receptors, to intracellular effectors that control critical functions in human cancer cells, such as differentiation, growth, angiogenesis, and inhibition of cell death and apoptosis." Christofferson and colleagues (2009) argue, "The rationale for such therapy is the realization that, in general, oncogenes and tumor suppressor genes encode proteins that are mutated or dysregulated forms of key components in major regulatory pathways."

As promising as this sounds, there have been limited victories to date. Levitzki and Klein (2010) report, "With the notable exception of Gleevec for the treatment of early chronic myelogenous leukemia, targeted therapies have so far had limited success. Most cancers are not dependent on a single survival factor. Furthermore, tumors are constantly evolving entities, and are heterogenous in their cellular makeup, compounding the challenge."

Thus, we see yet another complex adaptive system as discussed in Chapter 1. While human researchers and clinicians adapt to increased understanding of cancer to create new therapies, cancer also adapts, both naturally and perhaps in response to our interventions, as bacteria have adapted to our evolving set of antibiotics. This suggests that success will be fleeting and a constant challenge, as is the case for many of the examples elaborated in this book.

Example – Urban Oceanography

Urban oceanography (Bruno & Blumberg, 2004) is concerned with the interactions of oceans and coastal cities. The urban environments are typically shallow coastal areas and estuaries, where oceanographic and atmospheric conditions exhibit high spatial and temporal variability due to the influences of freshwater inflows, combined sewer overflows, tides, microclimate, and bottom and land topography, which is subject to change because of dredging/deepening programs. Understanding these environments is further complicated by the presence of high turbidity, strong stratification, strong currents, fog, vessel traffic and limited shoreline access. Of particular interest in the urban ocean are the impacts of hurricanes, monsoons, tsunamis, and even flooding due to heavy rains on estuaries, rivers, and bays adjacent to large urban populations. A major concern is being able to predict how water levels will affect primary urban infrastructures such as tunnels, subways, and roads, as well as delivery of electrical power, water, and food.

The Stevens Estuarine and Coastal Ocean Model (Blumberg & Mellor, 1987; Blumberg et al., 1999; Georgas & Blumberg, 2010) is a three-dimensional, free-surface, hydrostatic, primitive equation estuarine and coastal ocean circulation model – see Figure 4.2. Predicted variables include water level, 3D circulation fields (currents, temperature, salinity, density, viscosity, and diffusivity), significant wave height and period.

Its operational forecast application to the New York/New Jersey Harbor Estuary and surrounding waters began in 2006 (Bruno et al., 2006; Fan et al., 2006; Georgas et al., 2009a; Georgas, 2010), and includes forecasts of chromophoric dissolved organic matter and associated aquatic optical

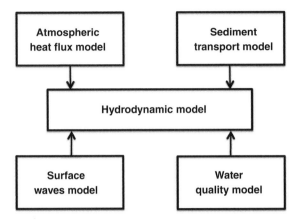

Figure 4.2   Urban Oceanography Model

properties through coupling to an water quality model (Georgas et al., 2009b). This system is called NYHOPS – New York Harbor Observing and Prediction System. The "observing" part of the acronym refers to real-time monitoring from water-borne sensors from Cape May to Cape Cod.

The central hydrodynamic model is a coupled set of partial differential equations. The model is computationally solved using finite difference methods that involve discretizing the continuous phenomena to a grid where solutions at each grid point are constrained by solutions at neighboring grid points. More specifically, the model computationally solves the hydrodynamic equations of motion and their attendant thermodynamic advection–diffusion equations. It incorporates a turbulent closure model that provides a realistic parameterization of vertical mixing processes, and a version of a horizontal mixing scheme for subgrid-scale horizontal shear dispersion. The model is forced in the open-ocean lateral boundaries by total water level, waves, and long-term thermohaline conditions, at the surface with a two-dimensional meteorological wind stress and heat flux submodel, and internally with thermodynamic inputs from river, stream, and water pollution control plant discharges, and thermal power plant recirculation cells. Quadratic friction is applied at the bottom based on internally calculated friction coefficients that include wave boundary layer effects, and at the free surface through assimilation of surface ice cover friction.

This model was highly successful in predicting the water levels that would result in greater New York City due to the impending Hurricane Sandy.

NYHOPS is used to produce daily 72-h predictions of water properties including water levels in the Hudson River, East River, New York/New Jersey Estuary, Raritan Bay, Long Island Sound, and the coastal waters of New Jersey. The US Coast Guard, and many others, in the marine community use these predictions because of their established accuracy.

DESIGNED PHENOMENA

Physical phenomena associated with the engineered world are usually designed to create desired functionality such as lifting and movement of people and goods, heating and cooling of physical spaces, and transformation of physical materials from one form to another. Such phenomena are designed, rather than naturally occurring, and hence they are usually decomposable into constituent components. In many cases, the dynamic responses of these components over time can be represented using lumped parameter models and ordinary differential equations. Such models are useful in electrical systems and electronics, mechanical systems, heat transfer, and acoustics.

Table 4.3 summarizes the central disciplines within engineering. The focus of each discipline is based on the definitions of Sections 1–10 of the National Academy of Engineering. The phenomena column in Table 4.3 reflects the expertise of members of each discipline. Thus, depending on the phenomena associated with the questions of interest, this column suggests what engineering disciplines need to be included on the team addressing these questions. Of course, for the types of problems discussed in Chapter 2, one would also need expertise in economics, finance, and the behavioral and social sciences.

Each of the engineering disciplines has a set of methods and tools that have been developed to address the phenomena associated with the discipline. There is an impressive range of approaches to modeling and representation of the "physics" of the environment, infrastructure, vehicles, and so on.

Models of the dynamic responses of systems are expressed in terms of stocks, flows, feedback, error, and control. These are expressed in terms of differential or difference equations. For the latter, there are continuous states, with discrete time transitions. Of particular interest are transient responses and, in particular, stability of responses. For example, these phenomena are very important in aircraft and other vehicles.

TABLE 4.3 Phenomena Associated with Engineering

Discipline	Focus	Phenomena
Aerospace engineering	Forces and physical properties of aircraft, rockets, flying craft, and spacecraft; aerodynamic characteristics and behaviors of airfoils and control surfaces, and properties such as lift and drag	Lift, drag, propulsion, flight, gravity, gas dynamics, stability, control
Biomedical engineering	Healthcare treatment, including diagnosis, monitoring, and therapy; Diagnostic and therapeutic medical devices, imaging equipment, regenerative tissue growth, pharmaceutical drugs, and therapeutic biologicals	Sensing, imaging, processing, detecting, diagnosing, dose–response relationships
Chemical engineering	Production, transformation, transportation, and proper usage of molecules, chemicals, materials, and energy; processes that convert raw materials or chemicals into more useful or valuable forms	Catalysis, chemical exchange, phase change and conversion, cracking, refining

Civil engineering	Design, construction, and maintenance of the physical and naturally built environment, including works like roads, bridges, canals, dams, and buildings; structures, transportation, earth, atmosphere, and water	Static and dynamic loading; flows of water, air, traffic, and so on; control of infrastructure processes
Computer science	Feasibility, structure, expression, and mechanization of the methodical processes (or algorithms) that underlie the acquisition, representation, processing, storage, communication of, and access to information	Information, logic, computation, representation, search, optimization
Computer engineering	Hardware and software aspects of computing; Design of electronics and software, integration of hardware and software, as well as integration of computers into larger systems	Electricity, electromagnetism, logic, computation, information, communications
Electrical engineering	Electricity, electronics, and electromagnetism for power, computation, communications, control, including signal processing; generation and transmission of electric power and use of electricity to process information	Electricity, electromagnetism, logic, computation, information, communications

(continued)

TABLE 4.3 *(Continued)*

Discipline	Focus	Phenomena
Industrial engineering	Development, improvement, implementation and evaluation of integrated systems of people, money, information, equipment, and so on; specification, prediction, and evaluation the outputs of such systems	Processes, networks, uncertainty, variability, queues, forecasting, planning, integration, optimization, decision making
Materials science	Application of the properties of matter to various areas of engineering; Relationship between the structure of materials at atomic or molecular scales and their macroscopic properties	Atomic and molecular structures; elasticity and plasticity; stress and strain properties; deformation
Mechanical engineering	Production and usage of heat and mechanical power to design, produce, and operate machines and tools; mechanics, kinematics, thermodynamics, materials science, structural analysis, and electricity	Mechanics, kinematics, and dynamics of solids, fluids and gases; thermodynamics and heat transfer

Another set of methods and tools are associated with discrete event systems characterized in terms of capacities, flows, queues, and allocations, all over time. Arrival and service processes are described probabilistically, for example, Poisson distributions of arrival time and exponential distributions of service times. For such systems, there are discrete states (e.g., numbers in queues) and continuous time transitions. Typically, the steady-state (versus transient) response is of interest in terms of average queue lengths, waiting time, and in-system times.

Another class of methods and tools is agent-based. Each element of the system is characterized in terms of information sampling rules and decision rules across perhaps thousands of agents. The phenomena of interest include how agents adapt to external information, how information is shared among agents, and the resulting overall population response. A primary goal is identification of emergent and often unexpected behaviors, as well as the triggers that lead to these behaviors.

For all three of these approaches, which represent a small subset of all engineering approaches, optimization is often an objective. For dynamic systems, one might want to determine the optimal feedback control laws that result in a desired balance between response error and energy or effort devoted to control. For discrete event systems, the allocation of resources (e.g., capacities), capacity routing and scheduling, and resource inventory levels are often the target of optimization, where performance versus resources required is typically the key trade-off. These possibilities are further pursued in Chapter 9.

In many situations, we are concerned with the physical impacts of the natural world on the designed world, as was illustrated by the earlier example of urban oceanography. The following examples focus on vehicle powertrains and manufacturing processes. At this point, we will not be concerned with the impacts of these systems on the environment. However, it is, of course, best to engineer designed phenomena so that they have beneficial interactions with natural phenomena. We return to this topic when discussing earth as a system in Chapter 7.

Example – Vehicle Powertrain

Figure 4.3 depicts a block diagram of a vehicle powertrain. This depiction is greatly simplified but sufficient to discuss central physical phenomena and how they are addressed. These phenomena include:

- *Combustion* leading to
- *Planar motion* of pistons in cylinders leading to

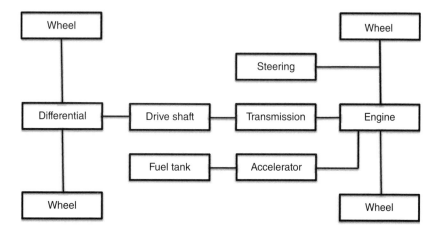

Figure 4.3 Vehicle Powertrain

- *Rotary motion* via crankshaft
- *Speed and torque conversion* via transmission
- *Differential turning speeds* of wheels via differential.

Understanding these phenomena requires expertise in three disciplines. Thermodynamics is concerned with heat and temperature and their relation to energy and work. In an internal combustion engine, combustion thermodynamics involves fuel, oxygen, and ignition via spark plugs.

Kinematics is concerned with the motion of points, objects, and systems of objects without consideration of the causes of motion. For the powertrain, kinematics is concerned with the conversion of planar motion to rotary motion via the crankshaft, speed and torque conversion via the transmission, and conversion of the drive shaft rotation into rotation of the rear two wheels at different speeds due to the differential.

Dynamics is concerned with the study of forces and torques and their effects on motion. For the powertrain, dynamics is concerned with conversion of explosive force to motion of masses in the powertrain including pistons, flywheel, drive shaft, axel, and wheels. The automobile driver is also central to the dynamics of the vehicle, but we will delay consideration of human manual control until Chapter 5.

Modeling and visualization of vehicle powertrains usually rely on dynamic system simulations that incorporate various computational representations drawn from thermodynamics, kinematics, and dynamics. The underlying mathematical formalisms employed by such simulations are discussed in Chapters 8 and 9.

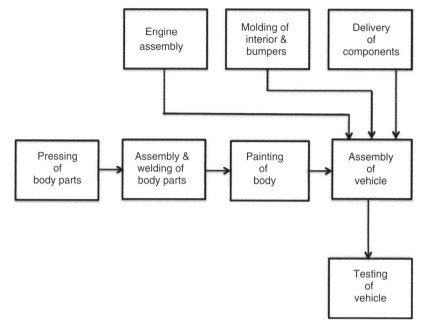

Figure 4.4   Vehicle Manufacturing

Example – Manufacturing Processes

Figure 4.4 depicts a block diagram of vehicle manufacturing. This depiction is also greatly simplified but sufficient to discuss central physical phenomena and how they are addressed. These phenomena include:

- Pressing and molding
- Welding
- Painting
- Part handling
- Assembly
- Testing.

Understanding these phenomena requires expertise in four disciplines. Strength of materials is concerned with the behavior of solid objects subject to stresses and strains. This expertise is needed to address the pressing and molding of body parts that will be assembled. This includes metal body parts as well as plastic bumpers and interior moldings and assemblies, for example, dashboards.

Metallurgy is concerned with production of metals and the engineering of metal components for use in products. This discipline is central to engine design and other cast components. Of course, strength of materials is also of great importance for products of metallurgical processes.

Paint chemistry is concerned with liquid or mastic compositions that, after application to a substrate in a thin layer, convert to solid films. Understanding of paint chemistry is central to the application of paint to vehicle body parts. The paint shop is a hugely expensive portion of vehicle assembly and also one of the least environmentally friendly.

Assembly is concerned with interchangeable parts that are added as the semifinished vehicle moves from station to station where the parts are added in sequence until the final vehicle is produced. The sequencing of assembly and layout of assembly lines are areas amenable to optimization, often with additional heuristics to handle contingencies not fully addressable by mathematics.

Vehicle assembly is usually represented in discrete-event simulations. Such simulations represent flows between stations and transformation at stations. Primary concerns are matching assembly capacities to flow characteristics to decreased assembly time per vehicle as well as errors and other quality issues.

DETERRING OR IDENTIFYING COUNTERFEIT PARTS

Counterfeiting has long been a problem of suppliers delivering parts of lesser quality than buyers had specified and expected. A primary motivation is lowering the costs of securing or producing the parts and thereby increasing profits. As Bodner and his colleagues (2013) indicate, "Counterfeit parts have different performance and failure characteristics than genuine parts and can result in degraded system availability, reliability and performance in the field, not to mention critical safety issues."

Figure 4.5 depicts a multilevel model of the context of counterfeiting. Omitting the overall economy and the defense acquisition governance systems still leaves an immensely complex system. There are an enormous number of stakeholders whose interests seldom align.

Bodner and colleagues further argue, "Concern centers around two types of counterfeiting – fraudulent counterfeits and malicious counterfeits. Fraudulent counterfeits derive from the traditional motivation of a counterfeiter to make a profit through fraud, by substituting an inferior product that is inexpensively produced relative to the cost of the genuine article. These types of counterfeits fall into several categories. First are parts that are re-marked

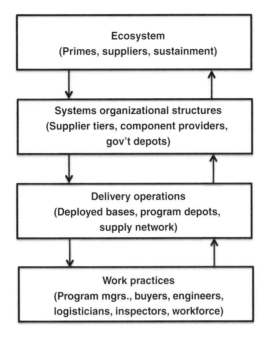

Figure 4.5 Multilevel Model of the Context of Counterfeiting

to appear that they are original equipment manufacturer parts. Second are defective parts that are passed as good original parts. Third are parts that are removed from scrapped assemblies and passed as new. Malicious counterfeits are designed to appear to perform correctly, but then malfunction at critical times or open security breaches so that adversaries gain advantage."

The complexity of supply chains and assembly of systems complicate deterrence of counterfeiting or identification of counterfeit parts. Complex systems are composed of major subsystems, which consist of lower level subsystems, assemblies, components, and so on. The problem is not that the overall system is counterfeit, but that some of its constituent elements may be counterfeit. The key question according to Bodner et al. (2013) is "how to detect counterfeit parts and prevent them from being installed in a system, or to detect counterfeits already installed."

They elaborate, "These constituent elements typically come from a variety of suppliers in a many-tiered supply network. A component may originally come from one supplier and pass through several others as it is installed in a sub-system, which is in turn installed in a major sub-system, and finally in an end-product system. Thus, identifying the source of counterfeits to prevent future counterfeit occurrences is not trivial. This is compounded by the globalization of the supply chains, especially in the area of electronic parts."

Finally, they note that "many systems are kept in use for decades, often past their expected lifetime, e.g., weapon systems, ships, road vehicles. Most systems consist of constituent parts specifically designed for that system or platform. As the years pass, the supply base is susceptible to supplier diminishment, whereby the original manufacturers of a component or sub-system exit the market, necessitating procurement of replacement parts for deployed systems from other sources. Finding these sources can be difficult."

There is a wide range of phenomena associated with this problem. Physical phenomena include fabrication and testing of parts. These parts then flow through physical supply chains to assembly at several levels. Finally, there is the physical functioning of the whole system once assembled. Overarching all of these phenomena is the systems engineering and management process.

Representation of these phenomena draws upon many engineering disciplines. Those associated with individual parts depend on their electrical, mechanical, material, or even chemical nature. Supply chains are usually represented using discrete-event simulations. The physical functioning of the fully assembled system is usually represented using various dynamic system modeling and simulation tools.

Using these methods and tools to represent the probabilistic nature of counterfeiting and the impacts on system functioning and performance is a daunting task due to the immense number of permutations of possibilities. Bodner and his colleagues (2013) illustrate a methodology for addressing this. Nevertheless, applying this methodology to a large-scale complex system would be a very substantial undertaking.

All of the aforementioned physical phenomena occur in a broader context of economic and social phenomena. There is the macroeconomics of procurement and operation of complex systems. This often includes substantial pressures on proposal prices, with the low bid frequently the winner. Lower prices require lower costs, which cascades throughout the supply networks.

The microeconomics of the firms within the supply networks are nested within the macroeconomic pressures. Investments in production efficiencies inevitably reach diminishing returns. If price and hence cost pressures persist, supplier firms may be incentivized to cut costs by using cheaper materials, short-circuiting testing, or recycling used parts that are difficult to identify.

Such behaviors can become socially acceptable when suppliers perceive associated procurement processes as unfair, especially when they share these perceptions with peer suppliers. Media attention to procurement issues can also contribute to a shared sense that everyone is forced to cheat the system.

Procurement policies and practices can affect the decisions made by supplier firms, not always positively. For example, price controls in healthcare delivery have resulted in behaviors by suppliers that are dysfunctional for the overall system but completely rational for the individual firm (Rouse & Serban, 2014). The key concern is finding the best mix of incentives and inhibitions to deter counterfeiting as well as penalize it once detected.

Bodner and colleagues review alternative strategies. Acquisition-oriented strategies include exclusive reliance on trusted suppliers, perhaps with emphasis on subsidiaries of prime system contractors. Analytic strategies could be employed to identify critical parts meriting extra attention. Software assurance methods and tools are another example. Design approaches include robust system design such that the system can still function with counterfeit components, perhaps with graceful degradation. Trusted system design involves creating capabilities for the system to detect and/or disallow counterfeits.

Sustainment-oriented strategies also include reliance on trusted suppliers, perhaps subsidiaries of primes as noted earlier. Supply chain monitoring can be used to prevent, detect, and respond to counterfeits, probably with incentives to prime and secondary contractors to perform such monitoring. Traceable components would help, with reporting and information sharing, perhaps laced with a bit of intelligence. Finally, of course, there would be penalties for counterfeiting, as well as for allowing counterfeits to be passed in subsystems or overall systems.

How might the types of models outlined earlier be used to evaluate the efficacy of these alternative strategies? Models of the physical flow of parts, their assembly into systems, and subsequent system performance could be used to evaluate and refine the notions of robust system design and trusted system design. If successful, these approaches would, in theory at least, enable systems to adapt to counterfeits.

Beyond the issue of feasibility, another major issue is the increased costs of design, acquisition, and sustainment. The resulting increased complexity of systems could have side effects such as an overall slowing of design and development processes, leading to longer times to field new capabilities. As a result, existing capabilities will have to remain in service longer.

Alternatively, macroeconomic and microeconomic models could be used to evaluate likely impacts of economic incentives and penalties. Thus, rather than investing in attempts to make systems impervious to counterfeits, the focus would be on deterrence. Bonuses for sustained delivery of high-quality parts as well as penalties for unacceptable parts could be evaluated with these models. Penalties might range from no payment for bad parts to loss of supply contract to banishment from the supply ecosystem.

The feasibility of these approaches depends on being able to detect counterfeits, which would require some form of economic sampling and testing strategy. To reduce costs, one could just focus on nontrusted suppliers, although there have been recent reports of counterfeits originating in wholly owned subsidiaries of prime contractors. The cutting of corners can happen in various ways for many reasons.

Penalizing of suppliers can run afoul of other priorities. For example, sales in a particular country may require local content, that is, a portion of the supply contracts being awarded to companies in that country. It is quite possible that there might be no trusted suppliers in that country. Thus, to gain the revenues from these sales, processes would be needed to evaluate suppliers as well as assure their sustained acceptable performance. This may be difficult in countries whose business practices differ substantially from those in the countries of the prime contractors.

Finally, Bodner and colleagues (2013) offer a caution, "Counterfeiters have the ability to adapt to new circumstances, such as policies or procedures designed to detect or prevent counterfeit parts, with new methods to pass counterfeits onto their targets. Thus, those who would combat counterfeiting must adapt, as well. Such adaptive behavior is a hallmark of (complex adaptive) systems."

It is easy to see these phenomena being analogs to our earlier discussions of biology and cancer.

CONCLUSIONS

This chapter considered two types of physical phenomena. First, we discussed naturally occurring phenomena such as weather and water flow. We considered two examples – human biology and urban oceanography. Then we addressed designed phenomena such as systems engineered to move people and goods. Examples of interest here include vehicle powertrains and manufacturing processes. Of course, we are often concerned with the intersection of designed and natural physical phenomena. The chapter concluded with an elaboration of the counterfeit parts problem.

REFERENCES

Bianco, R., Melisi, D., Ciardiello, F., & Tortora, G. (2006). Key cancer signal transduction pathways as therapeutic targets. *European Journal of Cancer*, 42, 290–294.

Blumberg, A. F., & Mellor, G.L. (1987). A description of a three-dimensional coastal ocean circulation model, in N.S. Heaps, Ed., *Three-Dimensional Coastal Ocean Models* (pp 1–16). Washington, DC: American Geophysical Union Washington.

Blumberg, A. F., Khan, L.A., & St. John, J.P. (1999). Three-dimensional hydrodynamic model of New York Harbor region. *Journal of Hydraulic Engineering.* 125, 799–816.

Bodner, D.A., Prasad, P., Sharma, V., Compagnoni, A., & Ramirez-Marquez, J.E. (2013). *A Socio-Technical Model of the Problem of Counterfeit Parts.* Hoboken, NJ: Systems Engineering Research Center.

Bruno, M.S., & Blumberg, A.F. (2004). An urban ocean observatory for the maritime community. Real-time assessments and forecasts of the New York harbor marine environment. *Sea Technology,* 45 (8), 27–30.

Bruno, M.S., Blumberg, A.F., & Herrington, T.O. (2006). The urban ocean observatory – coastal ocean observations and forecasting in the New York Bight. *Journal of Marine Science and Environment.* C4, 1–9.

Christoffeson, T., Guren, T.K., Spindler, K.G., Dahl, O. Lonning, P.E., Gjertsen, B.T. (2009). Cancer therapy targeted at cellular signal transduction mechanisms: Strategies, clinical results and unresolved issues. *European Journal of Pharmacology,* 625, 6–22.

Fan, S.A., Blumberg, A.F., Bruno, M.S., Kruger, D., & Fullerton, B. (2006). The skill of an urban ocean forecast system. *Proceedings of 9th International Conference in Estuarine and Coastal Modeling.* Charleston, South Carolina, October 31–November 2, 2005, 603–618.

Georgas, N., (2010). *Establishing Confidence in Marine Forecast Systems: The Design of a High Fidelity Marine Forecast Model for the NY/NJ Harbor Estuary and its Adjoining Waters.* Hoboken, NJ: Stevens Institute of Technology, PhD. Dissertation.

Georgas N., & Blumberg, A.F. (2010). Establishing confidence in marine forecast systems: The design and skill assessment of the New York harbor observation and prediction system, *Proceedings of 11th International Conference in Estuarine and Coastal Modeling,* Seattle, Washington, November 4–6, 2009, 660–685.

Georgas, N., Blumberg, A.F., Bruno, M.S., & Runnels, D.S. (2009a). Marine Forecasting for the New York urban waters and harbor approaches: The design and automation of NYHOPS. *Proceedings of 3rd International Conference on Experiments / Process / System Modelling / Simulation & Optimization.* 1, 345–352.

Georgas, N., Li, W., & Blumberg, A.F. (2009b). *Investigation of Coastal CDOM Distributions Using In-Situ and Remote Sensing Observations and a Predictive CDOM Fate and Transport Model.* Arlington, VA: Office of Naval Research, Ocean Battleship Sensing Science & Technology Program.

Hahn, W.C., & Weinberg, R.A. (2002). Rules for making human tumor cells. *New England Journal of Medicine,* 349, 1593–1603.

Ingber, D.E. (2003). Tensegrity I. Cell structure and hierarchical systems biology. *Journal of Cell Science,* 116, 1157–1173.

Levitzki, A., & Klein, S. (2010). Signal transduction therapy of cancer. *Molecular Aspects of Medicine*, 31, 287–329.

Milella, M., Ciuffreda, L., & Bria, E. (2010). Signal transduction pathways as therapeutic targets in cancer therapy, in L.H. Reddy & P. Couvreur, Eds., *Macromolecular Anticancer Therapies (Chapter 2)*. Berlin: Springer.

Mukherjee, S. (2010), *The Emperor of All Maladies: A Biography of Cancer*. New York: Scribner.

Pavé, A. (2006), Biological and ecological systems hierarchical organization, in D. Pumain, Ed., *Hierarchy in Natural and Social Science*. New York: Springer.

Rouse, W.B., & Serban, N. (2014). *Understanding and Managing the Complexity of Healthcare*. Cambridge, MA: MIT Press.

5

HUMAN PHENOMENA

Table 5.1 lists the human phenomena of interest for the six archetypal problems introduced in Chapter 2 and discussed in Chapter 3. All of these phenomena are concerned with human perceptions, expectations, decisions, and performance, either individually or in teams or groups. This chapter addresses the state of knowledge in representing these phenomena.

This chapter proceeds as follows. First, descriptive and prescriptive approaches are contrasted. Descriptive approaches focus on data from past instances of the phenomena of interest. Prescriptive approaches attempt to calculate what humans should do given the constraints within which they have to operate.

The subsequent section outlines a wide range of models of human behavior and performance. There are numerous compilations and textbooks to draw upon. The relative emphases and levels of treatments in each of these sources are elaborated. This exposition is followed by a detailed look at the traffic control problem from our set of archetypal problems.

Many of the models discussed make assumptions about what humans know relative to the tasks at hand. Some of this knowledge is characterized using the notion of "mental models." The section following the detailed example addresses the nature of this construct in terms of approaches to assessing

Modeling and Visualization of Complex Systems and Enterprises:
Explorations of Physical, Human, Economic, and Social Phenomena, First Edition. William B. Rouse.
© 2015 John Wiley & Sons, Inc. Published 2015 by John Wiley & Sons, Inc.

TABLE 5.1   Human Phenomena Associated with Archetypal Problems

Problem	Phenomena	Category
Cancer biology	Human decisions on diagnoses given test results	Human, individual
Cancer biology	Human decisions to employ drugs targeted at signaling mechanisms	Human, individual
Financial system	Human decisions by investors relative to risks and rewards	Human, individual
Traffic control	Human decisions by individual drivers on where and when to drive	Human, individual
Traffic control	Human performance in manual control of vehicle	Human, individual
Urban resilience	Human perceptions, expectations, and intentions	Human, individual
Counterfeit parts	Human decisions by supplier executive teams	Human, team or group
Healthcare delivery	Human decisions of clinical teams on use of capacities	Human, team or group
Traffic control	Human decisions by traffic management teams on pricing	Human, team or group
Urban resilience	Human decisions of urban teams on communications	Human, team or group

mental models and use of the outcomes of such assessments. The final section of this chapter addresses fundamental limits in modeling human behavior and performance.

DESCRIPTIVE VERSUS PRESCRIPTIVE APPROACHES

There are many interpretations of the word "model." There are fashion models, this year's car models, and toy models for children and occasionally adults. The meaning in this chapter is much more focused. A model is a mechanism for predicting human behavior and performance for specific tasks in particular contexts.

One approach to such modeling involves using empirical data previously collected that is applicable to the tasks and contexts of interest. Thus, one might, for example, have data on visual acuity as a function of character font and size, as well as ambient lighting. This might take the form of several equations relating probability of reading errors to character size and lighting level for each font type. One would assume that these relationships hold for the new task and context, and then interpolate and carefully extrapolate to predict reading errors in the new situation.

This approach to modeling is termed "descriptive" in the sense that one describes past behavior and performance, assumes these data apply to the current tasks and context, and interpolates or extrapolates to make predictions. If effect, one is saying that the past data describe how humans performed and such descriptions provide good predictions of how humans will perform in the new situation.

This works well for aspects of behaviors and performance that are not too context sensitive. Human abilities to see displays, reach controls, and exert force are good examples. On the other hand, this approach is much less applicable to phenomena compiled in Table 5.1. The limitation of the descriptive approach is the difficulty of compiling the many task- and context-specific empirical relationships in advance of encountering new design issues and problems.

The "prescriptive" approach focuses on how humans should behave and perform. This, of course, begs the question of whether people can conform to such prescriptions. This reality led to the notion of "constrained optimality." Succinctly, it is assumed that people will do their best to achieve task objectives within their constraints such as limited visual acuity, reaction time delays, and neuromotor lags. Thus, predicted behaviors and performance are calculated as the optimal behavior and performance subject to the constraints limiting the humans involved. Typically, if these predictions do not compare favorably with subsequent empirical measurements of behaviors and performance, one or more constraints have been missed (Rouse, 1980, 2007).

Determining the optimal solution for any particular task or tasks requires assumptions beyond the likely constraints on human behavior and performance. Many tasks require understanding of the objectives or desired outcomes and inputs to accomplish these outcomes, as well as any intervening processes. For example, drivers need to understand the dynamics of their vehicles. Pilots need to understand the dynamics of their aircraft. Process plant operators need to understand the dynamics of their processes. They also need to understand any trade-offs between, for example, accuracy of performance and energy expended, human and otherwise, to achieve performance.

This understanding is often characterized as humans' "mental models" of their tasks and context. To calculate the optimal control of a system, or the optimal detection of failures, or the optimal diagnoses of failures, assumptions are needed regarding humans' mental models. If we assume well-trained humans who agree with and understand task objectives, we can usually argue that their mental models are accurate, for example, reflect the actual physical dynamics of the vehicle.

For novice humans, this is an unwarranted assumption. For humans that may or may not agree with task objectives, this assumption may also be indefensible. For these and other reasons, models of social phenomena are a greater challenge, as elaborated in Chapter 7. However, within the realm of highly trained professionals operating within designed systems, the prescriptions of constrained optimality have been rather successful.

MODELS OF HUMAN BEHAVIOR AND PERFORMANCE

There are many expositions of models of human behavior and performance that can be drawn upon when concerned with human phenomena. There have been two studies by the National Academies and numerous books published on this topic.

Pew and Mavor's (1998) National Academy study reviewed models of human behavior and performance in terms of attention and multitasking, memory and learning, human decision making, situation awareness, planning and behavior moderators, as well as integrative architectures. They acknowledged the richness of this knowledge base, but also questioned the maturity of the knowledge relative to the recognized modeling needs. One can argue, however, that modeling policy decision makers' behaviors by starting with how they visually read characters on printed pages or electronic screens will be more overwhelming than useful. It is essential that human phenomena be addressed at levels of abstraction compatible with the problems of interest.

More recently, Zacharias and his colleagues' (2008) National Academy study considered different levels of models, including verbal conceptual models, cultural modeling, macrolevel formal models (e.g., systems dynamics models and organizational modeling), mesolevel formal models (e.g., voting and social choice models, social network models, and agent-based modeling), microlevel formal models (e.g., cognitive architectures, expert systems, decision theory, game theory, and interactive games). This range of modeling methods and tools is much more relevant to the phenomena discussed in this chapter relative to the earlier Academy study.

Sheridan and Ferrell's (1974) classic on human–machine systems addresses modeling of a wide range of human behavior and performance. Humans' abilities to deal with uncertainty are characterized in terms of probability estimation, Bayesian probability revision, information measurement and channels, information transmission tasks, and continuous information channels. They provide an in-depth review of manual control performance including servomechanism models, input–output identification in time and frequency domains, and optimal control models. They summarize human characteristics in terms of sensory and neuromuscular abilities, as well as intermittent and nonlinear characteristics. Human decision making and utility, decisions under risk, signal detection, dynamic decision making, and formal games are discussed. The knowledge base in this classic is quite rich, although it is mostly focused on individual behavior and performance.

Example – Manual Control

Manual control, in contrast to automatic control, involves a human generating a control action in an attempt to cause the state of a process to assume desirable values. As shown in Figure 5.1, the human is assumed to observe a noisy version of the difference between the desired output and the actual output of the process. The noise is added to reflect the fact that the human is unlikely to observe or measure this difference exactly. The input to the process is also assumed to be noisy due to neuromotor noise when moving the controls.

For the task depicted in Figure 5.1, we can view the human as an error-nulling device. In control theory terms, we can say that the human acts as a servomechanism. Using this servomechanism analogy, we can proceed to analyze the human-process system as if the human were in fact a servomechanism. This enables employing powerful modeling methods from control theory, as well as estimation theory – see Chapter 9.

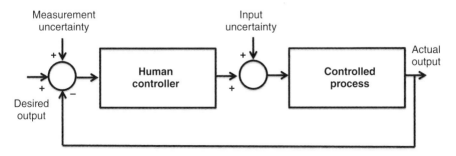

Figure 5.1 Block Diagram of Manual Control

Manual control has a rich history. Progressively more sophisticated problems of display and control design have been addressed. For example, what if the human was provided with a preview of the future desired output, just as automobile drivers can see the road ahead? What if the human was provided with a display of the predicted actual output to compare to the preview of future desired output? Intuitively, one would think that these two types of information would improve performance. Manual control models have been developed to predict the extent of improvement and empirical human-in-the-loop studies have supported these predictions.

Sheridan (1992) addresses supervisory control where humans interact with complex systems via computers rather than directly, which is quite common in most systems now. He outlines the generic supervisory control functions of planning, teaching the computer, monitoring automatic control, intervening to update instructions or assume direct control, and learning from experience. He discusses extensions of manual, as opposed to automatic, control theory beyond his earlier treatment. He reviews contemporary results on human attention allocation models, fuzzy logic models, and cognition and mental models. He concludes by discussing limiting factors – free will, ambiguity, and complexity – that make prediction of human behavior and performance challenging. These factors are central to several of the discussions in this book, particularly in Chapter 7.

Rouse (1980) presents a wide range of systems engineering models of human–machine interaction. Estimation theory models for state estimation, parameter estimation, and failure detection are discussed. Control theory models for manual control; quickening, prediction, and preview displays; and supervisory control are reviewed. Queuing theory models of visual sampling, monitoring behavior, and attention allocation are illustrated. Fuzzy set theory models for process control and fault diagnosis are discussed. Finally, artificial intelligence models are presented in terms of production systems, pattern recognition, Markov chains, and planning models. Overall, he shows how the "hard" methods of system dynamics and control, as well as operations research, can be applied to modeling human behavior and performance.

Rouse (1983) summarizes a wide range of models of human problem solving in the tasks of failure detection, failure diagnosis, and failure compensation. He reviews eight mathematical models of failure detection and eleven mathematical models of failure diagnosis. The key conclusion is that there is a rich base of computational models to draw upon for modeling human behavior and performance in detection and diagnosis tasks.

TABLE 5.2   Problem-Solving Decisions and Responses

	Decision	State-Oriented Response	Structure-Oriented response
Recognition and classification	Frame available?	Invoke frame	Use analogy and/or basic principles
Planning	Script available?	Invoke script	Formulate plan
Execution and monitoring	Pattern familiar?	Apply appropriate S-Rules	Apply appropriate T-Rules

Example – Problem Solving

Drawing upon a wide range of sources (Rasmussen & Rouse, 1981), Rouse presents a general three-level representation of human problem solving (Rouse, 1983). Rasmussen's distinctions among skill-based, rule-based, and knowledge-based behaviors (Rasmussen, 1983), in combination with Newell and Simon's (1972) theory of human problem solving, led to the conclusion that problem solving occurs on more than one level – see Table 5.2.

When humans encounter a decision-making or problem-solving situation, they first consider available information on the state of the system. If this information maps to a familiar pattern, whether normal or abnormal, they perhaps unconsciously invoke a frame (Minsky, 1975) associated with this pattern. This enables them to activate scripts (Schank & Abelson, 1977) that enable them to act, perhaps immediately, via symptomatic rules (S-Rules) that guide their behaviors.

If the observed pattern of state information does not map to a familiar pattern, humans must resort to conscious problem solving and planning (Johannsen & Rouse, 1983), perhaps via analogies or even basic principles. Based on the structure of the problem, which typically involves much more than solely observed state variables, they formulate a plan of action and then execute the plan via topographic rules (T-Rules). As this process proceeds, they may encounter familiar patterns at a deeper level of the problem and revert to relevant S-Rules.

This framework has important implications for multilevel modeling of complex systems laced with behavioral and social phenomena. Succinctly, it may not make sense to represent human behavior and performance for any particular task using only one type of model. Scripted behaviors may be reasonable for familiar and frequent instances of these tasks. However, for unfamiliar and/or infrequent instances of these tasks, a more robust representation is likely to be needed.

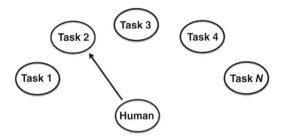

Figure 5.2   Multitask Decision Making

Rouse (2007) presents an expanded and updated set of the foregoing models, all premised on the nature of human abilities and limitations that people bring to their tasks. He elaborates estimation, queuing, control, and diagnosis models. He also provides models of human behavior and performance in system design, information seeking, multistakeholder decision making, investment decision making, strategic management, and enterprise transformation. The overall exposition addresses human behavior and performance in tasks ranging from operation and maintenance of complex systems to managing enterprises and leading them through fundamental change.

Example – Multitask Decision Making

Figure 5.2 depicts a situation where a human has to perform multiple tasks. For example, a human driving a car has several tasks – lateral control (steering), longitudinal control (accelerating/decelerating), scanning instruments, and so on. Another example is the aircraft pilot whose tasks include lateral control, longitudinal control, communicating with air traffic control, checking and updating radio fixes, monitoring the aircraft's subsystems, and so on. Yet another example is the human who monitors multiple processes in a power plant of a chemical plant.

In most situations, not all of the tasks require attention simultaneously. In other words, tasks "arrive" at different times. Many tasks, such as emergencies, arrive randomly. Some tasks are, of course, scheduled. However, if the schedule is very loose, such as aircraft arriving at an airport, the flow of arrivals is quite likely to appear random. The human has to decide which arrivals need attention when. High priority tasks may need immediate attention, while lower priority tasks can be performed when time is available.

Queuing theory (see Chapter 9) has been used to model human decision making in multitask situations. Various priority queuing models have been employed and empirically validated for complex multitask environments like flight management. A particularly interesting result is for Poisson distributed task interarrival times (i.e., randomly arriving tasks) with arrival rate λ_k for class k tasks, arbitrary distributions of service times with service rate μ_k for class k tasks, and cost of delay c_k for class k tasks. For this priority queuing model, the optimal servicing of tasks should be in order of decreasing $\mu_k c_k$.

It is interesting to consider why the arrival rate for each class of tasks (λ_k) does not affect priorities. For example, consider a situation where there are two classes of tasks and c_1 is much greater than c_2. Suppose there are no class 1 tasks waiting for service but there are class 2 tasks waiting. Perhaps one should wait for a class 1 task to arrive because c_1 is so much greater than c_2. Thus perhaps one should wait a short time Δt in hopes of a class 1 task arriving. If a class 1 task does arrive during Δt, then one would naturally service it. However, what if there are no class 1 arrivals during Δt? Then because of the lack of memory of the Poisson process, one is faced with exactly the same decision problem. To be consistent, one must choose to wait Δt again. In this way, one would continually wait for class 1 arrivals and never service class 2 tasks. Thus, the average waiting time would be infinite. From these arguments, one can see why arrival rates do not enter into the decision criterion.

Rasmussen (1986) and Rasmussen et al. (1994) discusses a range of models for attention allocation, signal detection, manual control, and decision making. With regard to decision making, he addresses human judgment, decision theory, behavioral decision theory, psychological decision theory, social judgment theory, information integration theory, attribution theory, fuzzy set theory, scripts, plans and expert systems, and problem-solving models. Rasmussen also presents three important conceptual frameworks: the means-ends abstraction hierarchy, levels of human control, and human error mechanisms. These frameworks have significantly influenced the lines of reasoning in this book.

Carley (2002a, b) and Carley & Frantz (2009) addresses computational modeling of socio-technical systems. She represents these systems as "synthetic agents composed of other complex, computational and adaptive agents constrained and enabled by their position in a social and knowledge web of affiliations linking agents, knowledge and tasks." She argues that the capabilities of agents (cognitive, communications, information seeking) define what types of "social" behaviors emerge, and concludes that, "The use of computational models enables generation of meaningful insights and the evaluation of policies and technologies."

Carley and Frantz (2009) discuss a set of computational tools for simulation of social systems. They build on a meta-matrix representation of who is connected to who and the nature of the connections. Their methods and tools include DyNetML, a universal data exchange format for social network data; Automap, a software tool for extracting semantic networks and meta-networks from raw, free-flowing, text-based documents; Organization Risk Analyzer, a software tool that computes social network, dynamic network, and link analysis metrics on single and meta-network data; and CONSTRUCT, a software tool that provides a platform to support virtual experimentation with meta-matrix data.

Barjis (2011), Carley (2002b), Cioffi-Revilla (2010), and Dietz (2006) discuss a variety of methodological considerations related to the characteristics of human and natural systems as well as the defining features of simulation models. They elaborate the notions of enterprise ontologies, enterprise governance, and enterprise architecture. This material represents recommended ways of thinking about modeling complex systems more than presenting models per se.

TRAFFIC CONTROL VIA CONGESTION PRICING

It is useful to consider how the rich set of models just outlined can be applied to one of the archetypal problems, in this case the traffic control problem. Human behavioral and social phenomena are manifested at several levels. Individual humans drive automobiles to destinations of their choice, making vehicle control decisions all along the way. Individuals interact with other drivers and their vehicles in traffic and at stop signs and lights.

Individuals' choices of destinations, for example, home in the suburbs, are often choices made by both the driver and family members. The routes they chose, for example, from work to home, are influenced by habits, traffic conditions, and other intentions such as stopping for groceries. These influencing factors may change during the course of driving.

The routes available are influenced by other humans, based on predictions of where traffic will flow, perhaps due to new office buildings and subdivisions. Other humans authorize and appropriate funds for road construction, typically at a state level, as influenced by other human decisions at the federal level.

All of the above occur in the context of social norms and values formed over years or decades. Many people value a home with a yard and swing set, but an apartment next to a large city park is also valued by many others. Media, entertainment, and advertising can influence the evolution of values and norms.

People at all levels are constantly adapting to the changing environment. A new office park opens and drivers quickly find a shortcut through the park that avoids a frustrating light. A couple becomes "empty nesters" and moves to a condominium next to the city park. Increased fuel efficiency decreases gas tax revenues and, therefore, funds for maintenance of transportation infrastructure.

Some adaptations occur quickly, such as spotting the new shortcut. Others take years like plans to move downtown and increasing the efficiency of cars. Changes of values and norms may take decades. These adaptations, at their respective time scales, are all happening simultaneously.

Consider the design of a new transportation network for an area that is rapidly urbanizing due to major new employers locating near recently discovered natural resources. This network design project includes major roads, traffic signaling, and congestion pricing, as well as light rail and busses for public transportation. Minor roads and side streets either already exist or will be created by developers of subdivisions, shopping centers, and entertainment venues.

How might the aforementioned levels of human behavior be represented in the engineering models developed in conjunction with this project? At the level of people driving cars, there are well-developed models of human manual control that provide reasonable predictions of lateral and longitudinal tracking errors. There are also models of human behaviors in traffic, but driver-to-driver interactions are challenging.

There are also models of drivers' choices among alternative routes. Key elements of such models are assumptions about what the driver knows about alternative routes, pricing if relevant, traffic conditions, and even availability of parking spots at the intended destination. With the advent of driver aids based on the Global Positioning System (GPS), including smart phones, there is potentially a wealth of information available to every driver. Appropriate assumptions about how people do or do not use this information are central to making reasonable traffic predictions.

Traffic projections are central to the design of road networks. One could design an initial network, simulate traffic flow using the driver choice models, and refine the design. This would require assumptions about where people travel from and to, which will change with the development of new subdivisions, shopping centers, and entertainment venues. TRANSIMS is a good example of such modeling, in this instance for Albuquerque (Nagel et al., 1999).

Integration of dynamic congestion pricing will affect drivers' choices of routes and shift traffic patterns. Predicting the impacts would require understanding people's utility functions for convenience (or time) and money. Rather than predicting responses, one could begin with low pricing and slowly increase pricing to see the ranges at which traffic patterns shift. In other words, one could approach this experimentally rather than axiomatically.

At the next level, concern shifts to the economics and politics of transportation systems and construction. Macroeconomic models, perhaps with microeconomic models of the firms involved, could apply here. However, purely economic and technical perspectives may not suffice with stakeholders focused on organizational, political, and legal perspectives. As a result, one may need a very complicated multistakeholder, multiattribute utility function that would be very difficult to operationalize.

Finally, we have to consider the impacts of social values and norms. Attitudes about freedom of the roads, government intervention, traffic delays, and payment for access might be addressable by surveys and focus groups. From a modeling point of view, these phenomena might function like force fields rather than lumped input–output relationships.

How well can we address the human phenomena associated with the traffic control problem? Several types of model-based predictions can be done fairly well:

- Drivers' manual control behaviors can be predicted reasonably well because people have to conform to the dynamic constraints of the task, within the limitations of visual acuity, reaction time delays, and neuromotor lags.
- The time required to troubleshoot sources of car malfunctions can be predicted pretty well based on the functionality and topography of the car, assuming of course that a knowledgeable person is doing the troubleshooting.
- Given origins, destinations, and road network, we could reasonably predict routes chosen. Traffic expectations will affect such choices. Predictions of these expectations can be incorporated into route selection models.

All of these models involve representing how humans conform to the objectives and constraints of engineered systems – vehicle dynamics, powertrain functionality, and road networks. Constrained optimality can reasonably be assumed in the sense that humans will perform as well as they can subject to their physical and cognitive constraints (Rouse, 1980, 2007).

Successful predictions, therefore, depend on two things. First, the engineered system needs to be well understood. If the system is emergent rather than engineered, then behavior-shaping objectives and constraints need to be well understood. In the absence of this, it is difficult to predict human behavior.

Second, humans' objectives and constraints need to be well understood. It has to be reasonable to assume that people will try to do the best they can within these constraints. In contrast, if objectives are ambiguous or people are "satisficing" rather than optimizing, it is difficult to predict human behavior.

Both macroeconomic and microeconomic models are usually premised on strong assumptions about "economic man" pursuing maximum economic returns. For example, utility theory models of markets assume particular preference trade-offs and risk appetites. Models of the firm assume profit maximization.

These models are much more useful for predicting aggregate behaviors. Predicting a particular driver's trade-off between convenience and cost would be difficult. Predicting a population's overall tendencies might be more reasonable, although economists' predictions are often off target.

At the political and social levels, behaviors are dominated by human–human interactions. Organizational, political, and legal perspectives usually drive these interactions. Economic or technical optimization seldom governs outcomes. Instead, negotiations and deals prevail. Such outcomes are difficult to predict.

Thus, we are best prepared to predict human behaviors when they directly interact with an engineered system such as a car or roadway. As behaviors are less directly constrained by the engineered system, prediction becomes more difficult. We could have predicted driver performance in their new 1957 Chevrolet (Swift, 2014), but not their decision to move to the suburbs because of President Eisenhower's new interstate highway.

MENTAL MODELS

The notion of mental models was discussed earlier in this chapter. This ambitious construct embraces a broad range of behavioral phenomena. It also has prompted various controversies. On one hand, it would seem that people must have mental models of their cars, for example. Otherwise, how would people be able to so proficiently negotiate traffic, park their cars, and so on?

On the other hand, perhaps people have just stored in their memories a large repertoire of patterns of their car's input–output characteristics, a large look-up table, if you will, of steering wheel angles and vehicle responses. From this perspective, there is no "model" per se – nothing computational that derives steering wheel angle from desired position of the car.

This is a difficult argument to resolve if one needs proof of the representation of the model in one's head. Are their differential equations, neural nets, or

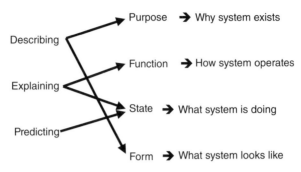

Figure 5.3   Functions of Mental Models

rules in the driver's head? Alternatively, one might adopt a functional point of view and simply claim that humans act as if they have certain forms of model in their brains that enable particular classes of behaviors.

We became deeply engaged in this issue, reviewing a large number of previous studies and publishing the often cited, "On Looking Into the Black Box: Prospects and Limits in the Search for Mental Models" (Rouse & Morris, 1986). We addressed the basic questions of what do we know and what can we know.

The definition that emerged, summarized by Figure 5.3, was, "Mental models are the mechanisms whereby humans are able to generate descriptions of system purpose and form, explanations of system functioning and observed system states, and predictions of future system states." This definition only defines the function of mental models, not what they look like.

We reviewed several methods for identifying mental models. Perhaps the most ubiquitous approach involves inferring characteristics of mental models via empirical study. This approach focuses on identifying the effects of mental models but not their forms. Thus, for example, our study of interpolation methods for displays of time series concluded that people's model of time series gives the same credence to the interpolated portions of the time series as the data points on which the interpolation is based. This conclusion does not invoke any specific form of model.

Another approach to identification is input–output modeling, effectively "fitting" a model to the relationship between displayed inputs and observed outputs. Thus, in the context of our studies of air traffic control, we might perform a regression of predicted future points versus displayed past points. This would tell us how people weight more recent points versus less recent points, for example. We would be, in effect, assuming that people's mental models look like regression equations.

Yet another approach is analytical modeling whereby we derive a mental model from first principles, use this model to predict human behaviors, measure actual behaviors in the same conditions, and then compare predictions with measurements. Our stochastic estimation model of human performance in estimation tasks (Rouse, 1977) is based on this approach. As with most analytic models, it does have a couple of free parameters, that is, fading memory rates, that can be used to adjust the model to better match human behaviors. However, these parameters are usually much fewer in number than the degrees of freedom of the behaviors being modeled.

Another approach to identification utilizes verbal and/or written reports by the humans whose behaviors are being studied, for example, (Ericsson & Simon, 1984). The value of this approach varies with the phenomenon being studied. People may be able to tell you how they think about stock prices, for example, but are much less able to tell you about how they think about riding a bicycle or driving a car, or even how they think about personal disputes. For this reason, verbal and/or written reports are often used to supplement other approaches to identification.

There are many issues associated with identifying mental models. An overriding issue is accessibility. Tasks that require explicit use of the functions depicted in Figure 5.3 are more likely to involve mental models that are accessible to the various methods just discussed. In contrast, tasks for which these functions are deeply below the surface of behavior are less likely to involve accessible models.

Another issue is the form of representation. Equations, neural networks, and rules are common choices. It is rarely possible to argue, however, that these representations actually reside in peoples' brains. Contemporary brain research might argue for neural networks being the substrate for all alternative representations. However, such an assertion hardly constitutes proof. We return to this issue later in this chapter.

Context of representation is another issue. To what extent are mental models general versus context specific? Do you have different models for one car versus another, or just one model with two sets of parameters? People seem, for instance, to have general rules of algebra rather than having a long repertoire of solutions to specific linear combinations of symbols and numbers. We seem much better at learning rules for solving problems rather than memorizing specific solutions. Perhaps, however, we are fooling ourselves in thinking about how we think.

A particularly interesting issue surrounds cue utilization. What do people actually pay attention to when solving problems and making decisions? To illustrate, recent studies have shown that people judge beauty in terms of the symmetry of faces. However, we are not aware of thinking of attractive people in terms of their symmetry – "Oh, she's so symmetric!" Thus, we do not always know what we are taking into account in performing our tasks.

All of the above come together when we consider the issue of instruction. What mental models should we attempt to create and how should they be fostered? Do people need to know theories, fundamentals, and principles, or can they just learn practices? The significance of this issue cannot be overstated.

To illustrate this point, I was involved in a dispute many years ago involving a curriculum committee of an engineering school. (I should preface this example with Henry Kissinger's well-known observation that the intensity with which faculty debate issues is only exceeded by the insignificance of the issues.) The committee was trying to agree, with little progress, on the required courses for the Ph.D. degree.

Given that we could not reach agreement, several faculty members argued that we should require more mathematics because "more math is always good for students." I asked these faculty members for empirical proof of this assertion. They reacted quite negatively to this request. I suggested that they were experts at mathematics but not education, having never studied or researched education per se. The senior-most faculty member stormed out of the room.

The question of what one needs to know is quite complicated and laced with subtleties. Contrary to most faculty members' perceptions – it should be noted that I have served on the faculties of five universities – students do not necessarily need to know what the faculty members know in order to succeed in their chosen paths in life.

A central element of this exploration of mental models concerns fundamental limits to what we can know. Several factors influence these limits. First, there are the mental models of researchers. Their methodological expertise, as well as the sociocultural norms within their context of research, has enormous impacts on how they address the mental models construct – or any research for that matter. If their subdiscipline requires, for example, convergence proofs of parameter estimates, their models of humans' mental models will be chosen to allow such proofs. Thus, the lenses of the investigators have an enormous impact (Kuhn, 1962; Ziman, 1968).

Another aspect of fundamental limits relates to the fact that inferential methods work best when humans have limited discretion in their task behaviors. If humans must conform to the demands of the environment to succeed in the task of interest, then our knowledge of the structure of the environment can provide a good basis for predicting behaviors.

On the other hand, if people have discretion – rather than predict the next point in the time series, they can escape to Starbucks – then we cannot solely rely on the structure of the task to predict their behaviors. This raises the obvious question of whether studies of mental models are really studies of the impacts of the structure of the environment. Herbert Simon has convincingly made this point (Simon, 1969).

We are also limited by the fact that verbalization works best when mental model manipulation is an inherent element of the task of interest. Troubleshooting, computer programming, and mathematics are good examples of tasks where mental model manipulation is central and explicit. In contrast, the vast majority of tasks do not involve explicit manipulation of task representations. Thus, our access of mental models – and the access of people doing these tasks – is limited.

One can easily argue for a form of uncertainty principle. Heisenberg (1927) showed that physicists could not measure both position and velocity perfectly accurately at the quantum level. You can measure either position or velocity accurately, but not both, mainly because the observation process affects what you are observing.

Translating this principle to the search for mental models, we cannot accurately assess both what a mental model is and what it is becoming because the act of assessment affects the model. For example, if I ask you to predict the future of the time series of stock prices, your ability to predict is affected because you now have to think about making such predictions.

Team Mental Models

We have thus far addressed individual behavior and performance. We now consider groups or teams, typically performing within a workspace where they can see or at least hear each other. We will consider both operational and management teams, as well as performing arts teams.

One of our studies of groups – actually teams – was prompted by the Aegis cruiser USS *Vincennes* shooting down an Iranian passenger airliner in the Persian Gulf on July 3, 1988 (Rogers et al., 1992). The Aegis weapon system was first commissioned in 1983 with the USS *Ticonderoga*. This system was developed to counter the serious air and missile threat that adversaries posed to US carrier battle groups and other task forces.

The *Vincennes* Incident prompted a congressional inquiry. Subsequently, the Office of Naval Research established a research program to study the potential behavioral and social factors underlying this incident. The TADMUS Program was named for its focus – tactical decision making under stress. I was the principal investigator for one subcontractor's efforts in this program.

We began by observing teams at the Aegis test facility in Moorestown, New Jersey. This training facility is unusual in that it looks like a ship super-structure rising out of a field west of the New Jersey Turnpike. It is sometimes referred to as the "Cornfield Cruiser."

Training exercises involved 25 people who staff the Combat Information Center (CIC). At the time of our observations, there were typically two exercises per day, each of which took considerable preparation, prebriefing, execution, and debriefing. I was struck by the fact that every exercise seemed to be aptly termed "25 guys face Armageddon." This is what Aegis is designed to do. However, as the July 3, 1988, incident shows, not all situations are Armageddon.

We focused our study on the anti-air activities of the CIC, as that is the portion of the team that dealt with the Iranian airliner. Our initial observation of this team suggested to us – Eduardo Salas, Jan Cannon-Bowers, and me – that team members did not have shared mental models. In particular, we hypothesized that inadequate shared models of teamwork – in contrast to mental models of equipment functioning or task work - hindered the team's performance (Rouse et al., 1992).

Our definition of mental models for teamwork followed the functional definition in Figure 5.3. However, the knowledge content, as shown in Table 5.3, differs from our earlier discussions. The overall research questions concerned what elements of Table 5.3 were needed by the anti-air team and how best to impart this knowledge.

The overall conclusions of this research were that teams were not well coordinated and did not communicate well in terms of their behaviors in these demanding exercises. It appeared that team members often did not know what was expected of them by other team members and did know what they could expect of others. Without expectations, they did not communicate or communicated erroneously or ambiguously. Occasionally, they could not explain or interpret communications received.

It was clear that these teams needed much-improved shared mental models of teamwork. This led to an intervention called Team Model Training that involved a desktop computer–based training system that approximated the full-scope Aegis simulator. An evaluation of Team Model Training compared this training with more conventional lecture-based training. Performance was assessed in the full-scope simulator subsequent to having received one of these training methods. Team Model Training significantly decreased the overall level of communications, indicating that well-trained teams are not necessarily those with the highest levels of explicit communications (Duncan et al. 1996).

TABLE 5.3   **Knowledge Content of Mental Models for Teamwork**

Level	Types of Knowledge		
	What	How	Why
Detailed/ specific/ concrete	Roles of team members (who member is)	Functioning of team members (how member performs)	Requirements fulfilled (why member is needed)
↓	Relationships among team members (who relates to who)	Co-functioning of team members (how members perform together)	Objectives supported (why team is needed)
Global/ general/ abstract	Temporal patterns of team performance (what typically happens)	Overall mechanisms of team performance (how performance is accomplished)	Behavioral principles/theories (why: psychology, management, etc.)

Performing Arts Teams

The research just discussed concerned training operational teams that typically are responsible for operations of ships, aircraft, process plants, factories, and so on. Another type of team is an organizational team, focused on managing projects, business functions, or whole enterprises. There is also a rich literature on this type of team, for example, Katzenbach and Smith (1993).

We undertook a study of a type of team that is a hybrid of operational and organizational teams – performing arts teams. Building upon the notion of mental models from Figure 5.3 and the knowledge content of team mental models in Table 5.3, we interviewed 12 performing arts leaders of symphony, chamber orchestra, chorus, jazz, musical theater, straight theater, improv theater, ballet, and puppetry. These interviews helped us to better understand the "ecology" of performing arts (Rouse & Rouse, 2004).

The interviews focused on the background of the arts leader and their organization, the nature of teamwork in their art form, the nature of any training in teamwork, either formal or informal, stories of particularly successful or unsuccessful instances of teamwork, and the role of the leader in their art form. Thus, our focus was on arts leaders' perceptions of the factors influencing teamwork and issues associated with teamwork.

Analysis of the interview data indicated that 5 dimensions were useful for differentiating the 12 performing arts organizations studied:

- Size of Performance – Number of performers and other participants
- Complexity of Performance – Extent of required coordination
- Locus of Coordination – Rehearsal versus performance
- Familiarity of Team Members - Ensemble versus pickup
- Role(s) of Leader – Prepares team; does versus does not perform.

Symphony orchestras and large choruses epitomize a large number of performers requiring extensive coordination with considerable rehearsal involving mostly ensemble performers and a leader who performs with the team. Opera and large musicals are similar with the primary exception that the leader seldom performs with the team. Jazz and improv theater are perhaps the other extreme, with a small number of performers with minimal coordination that is often accomplished during the performance. In this case, ensemble teams are the norm and leaders almost always participate in jazz and often participate in improv.

The nature of the performance interacts with these dimensions. The arts studied included music, words, and movement as the media of expression. Coordination to assure blending of performers' expressions is important in symphony orchestras and large choruses. This requires extensive rehearsal. Jazz and improv theater, in contrast, do not pursue blending in this sense. Spontaneity is central to these art forms. Preparation for these forms of performance does not tend to emphasize repetition. Ballet, for example, would seem to fall somewhere in between these extremes.

A question of particular interest is how the aforementioned dimensions affect teamwork and how teams are supported. Figure 5.4 provides a summary of potential relationships among these key variables as gleaned from the interview data. The primary outcome of interest is the extent of team training. Everyone interviewed extolled the benefits of teamwork; the prevalence of team training reflects explicit intentions to foster teamwork.

The solid arrows in Figure 5.4 designate crisp relationships, with upward-pointing deltas indicating positive relationships, and downward-pointing deltas indicating negative relationships. The dotted arrow designates a less-crisp relationship.

Not surprisingly, increasing team member familiarity decreases the prevalence of team training; further, increasing team training increases familiarity. Thus, ensemble teams may limit training to situations with new team members or possibly unusual productions. Selection may also be used to choose people who "fit in" and, at least stylistically, are therefore more familiar.

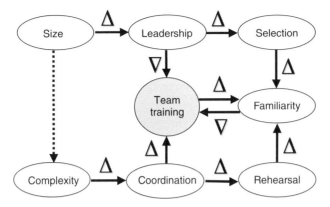

Figure 5.4 Relationships among Key Variables

The presence of strong leadership, especially leaders that perform, decreases the prevalence of team training. Such leadership also strongly affects selection, with the aforementioned impact on familiarity and, hence, team training. Rehearsal also increases familiarity. Needs for coordination strongly affect needs for rehearsal. Needs for coordination tend to increase with the complexity of the production.

Size affects needs for leadership, with the just noted impacts on the prevalence of team training and use of selection. Size and complexity, as indicated by the interview results, are not synonymous. Nevertheless, very large productions do tend to be more complex than small ones, at least very small productions. Except for these extremes, however, we expect that the correlation may be weak.

Note that the dynamics portrayed in this figure imply decreasing frequency of formal team training, due to either increasing familiarity or leadership decisions. High turnover among performing team members would tend to lower familiarity and, hence, increase use of team training until new performers are assimilated. Thus, team training may "come and go" with changing composition of performing teams.

Figure 5.4 provides a qualitative model or theory of the relationships among the dimensions of the ecology identified in terms of how these dimensions affect the prevalence of team training. This model does not predict how training will be pursued, or the impact of training on performance. Nevertheless, it suggests the situations where training will be employed to assure successful teamwork. The arts leaders interviewed perceived such teamwork to be central to successful performance.

Thus, it is clear that arts leaders recognize the importance of teamwork beyond taskwork. They see collaboration as central to excellence in the performing arts. The mechanisms that they adopt for assuring teamwork vary considerably with the nature of the art form, as well as for the reasons depicted in Figure 5.4.

The notion of mental models, while not explicitly suggested by any of the arts leaders interviewed, relates to two phenomena that were mentioned repeatedly. First, of course, performers need "models" of the performance at hand – what relates to what, when it relates, and, in some cases, why it relates. Second, performers need mechanisms to form appropriate expectations of fellow team members. In some cases, the score or script provides expectations, but in others, team members need deeper understanding of each other as individuals.

The means used to foster mental models vary considerably. The models associated with individual performance (i.e., taskwork) are assumed to be developed when team members are selected. Indeed, this tends to be an implicit or explicit selection criterion. In situations where the score or script does not fully define performance, additional means – often informal – are typically employed to enable people to understand each other's inclinations, motivations, and so on. This is common in theater and jazz, for example.

FUNDAMENTAL LIMITS

The investigation of limits in the search for mental models was the first of several related studies. This research led us to think about fundamental limits in general, for example, (Davis, 1965; Davis & Park, 1987; Glymour et al., 1987). We combed the literature for limits in physics, chemistry, biology, and other disciplines. It quickly became apparent that we are much better at understanding limits in axiomatic worlds (e.g., mathematics) than in natural worlds (e.g., the atmosphere).

These insights led us to look at limits to capturing humans' knowledge and skills (Rouse et al., 1989). We considered various representational forms in terms of variables, relationships among variables, and parameters within relationships. These forms ranged from probabilistic models to linear, nonlinear, and distributed signal processing models, to symbol processing models such as switching functions, automata, and Turing machines.

We explored cognitive modeling, wondering whether cognition may be like energy in that you can measure its impact but not the phenomena itself. We considered identification of both signal processing and symbol processing models in terms of uniqueness, simplicity, and acceptability of representations. These issues were pragmatically illustrated in the context of studies of three types of tasks: state estimation, air traffic control, and process control.

TABLE 5.4 Implications and Consequences of Modeling Limits

Implications of Modeling Limits	Consequences of Implications	
	Aiding	Automation
Inappropriate model	Advice wrong	Control wrong
Inadequate model	Advice incomplete	Control incomplete
Nonunique model	Misplaced compliance	Misplaced confidence

The ways in which these issues were addressed depended on the purpose of the modeling effort. These three studies were focused on enhancing human abilities and overcoming human limitations. This can be pursued via improved design, training, and/or automation. The effects of fundamental limits differ for these three avenues, as the next discussion illustrates.

We decided to undertake a study of fundamental limits of intelligent interfaces (Rouse & Hammer, 1991). Given that there are limits to modeling human behavior and performance, there must be limits on intelligent systems that rely on embedded models of human behavior and performance. Beyond understanding such limits, we also wanted to devise means for identifying when limits manifested themselves in specific intelligent systems.

There are several types of limits in model formulation:

- Data samples on which models are based are almost always limited
- Variables chosen for study may be incomplete and/or incorrect
- Structural assumptions may be inadequate and/or inappropriate
- Parameters estimates may be nonunique

Table 5.4 summarizes the implications of these types of limits. If flawed models are used to aid (or train) people, they will receive incorrect and/or incomplete guidance and may comply with this guidance. If such models are the basis for automation, then the automatic controller's behaviors will be incorrect and/or incomplete and humans will have misplaced confidence in the automation.

Our goal was to devise means for detecting, diagnosing, and compensating for the implications and consequences in Table 5.4. Detection can be based on examining input–output relationships. Three classes of events are of interest:

- Unanticipated output values, for example, out of range variables

- Missing output variables – a variable needed but not included
- Extra output variables – a variable included but not needed.

The modeling limitations and compensatory mechanism include:

- Model incorrect – modify structure
- Model incomplete – add relationships
- Model inaccurate – modify relationships
- Model incompatible – change representation.

Note that detection and diagnosis involve two overarching concerns. One is internal evaluation or *verification*. This addresses the question of whether the system was built as planned. The second concern is external evaluation or *validation*. This addresses the question of whether the plan solved the problem for which it was intended. The process we devised addresses both of these concerns.

We demonstrated this process by evaluating an intelligent information management system based on goal–plan graphs. These graphs relate a user's goals to their plans for achieving their goals to the information requirements for executing these plans. This intelligent information system was intended to infer the user's intentions – their goals – and then map to plans and information requirements, and subsequently automatically display the information needed by the user to pursue their goals.

Our detection procedure began with observed display discrepancies in terms of the three types of events listed earlier. Discrepancies were mapped to the information requirements that generated the display, and then to the plans and goals that drove these requirements, and finally to potential deficiencies in inferring intentions. This reverse chaining was the means for diagnosing the discrepancies in terms of the four types of modeling limits listed earlier.

We applied this methodology to evaluation of the Information Manager within the Pilot–Vehicle Interface of the Pilot's Associate, an artificially intelligent copilot for fighter aircraft. The goal–plan graph had been developed via "knowledge engineering" using a group of expert fighter pilots. The evaluation employed another group of pilots to assess any display discrepancies during a 27-min flight scenario. The detailed results are presented in Rouse and Hammer (1991).

Of particular relevance for this discussion, 22 modeling problems were identified in this 27-min flight. Only one of these problems was a "bug" in the sense of a programming error. The other 21 reflected both verification and validation problems. Compensation for these problems involved remediating incorrect, incomplete, inaccurate, and/or incompatible models.

Fundamental limits keep us from fully understanding human behavior and performance. Nevertheless, we gain what understanding we can and then we base designs, aiding, training, and automation on this understanding. As indicated in Table 5.4, there are important implications and consequences of this process. It is important that we develop and employ means for detecting, diagnosing, and compensating for these consequences.

CONCLUSIONS

This chapter has explored human phenomena, both individually and as groups or teams, in terms of behavior and performance – what people do and how well they do. There is a wealth of models, developed over 60+ years, to draw upon. To the extent that the environment constrains human behaviors, we can make reasonably good predictions of human behavior and performance. In contrast, when humans have considerable discretion in what they choose to do and how they behave, our abilities to predict suffer. We return to this issue in Chapters 7 and 10.

REFERENCES

Barjis, J. (2011). Enterprise modeling and simulation within enterprise engineering. *Journal of Enterprise Transformation*, 1, 185–207.

Carley, K.M. (2002a). Computational organization science: A new frontier. *Proceedings of the National Academy of Science*, 99 (3), 7257–7262.

Carley, K.M. (2002b). Computational organizational science and organizational engineering. *Simulation Modeling Practice and Theory*, 10, 253–269.

Carley, K.M., & Frantz, T.L. (2009). Modeling organizational and individual decision making. *Handbook of Systems Engineering and Management* (Chapter 18). Hoboken, NJ: John Wiley.

Cioffi-Revilla, C. (2010). A methodology for complex social simulations. *Journal of Artificial Societies and Social Simulation*, 13 (1), 7–25.

Davis, M., Ed. (1965). *The Undecidable: Basic Papers on Undecidable Propositions, Unsolvable Problems, and Computable Functions*. Hewlett, NY: Ravel Press.

Davis, P.J., & Park, D. (Eds.). (1987). *No Way: The Nature of the Impossible*. New York: Freeman.

Dietz, J. (2006). *Enterprise Ontology – Theory and Methodology*. Berlin: Springer-Verlag.

Duncan, P.C., Rouse, W.B., Johnston, J.H., Cannon-Bowers, J.A., Salas, E., & Burns, J.J. (1996). Training teams working in complex systems: A mental model based approach, in Rouse W.B., Ed., *Human/Technology Interaction in Complex Systems* (Vol. 8, pp. 173–231). Greenwich, CT: JAI Press.

Ericsson, K.A., & Simon, H.A. (1984). *Protocol Analysis: Verbal Reports as Data.* Cambridge, MA: MIT Press.

Glymour, C., Scheines, R., Spirtes, P., & Kelly, K. (1987). *Discovering Causal Structures: Artificial Intelligence, Philosophy of Science, and Statistical Modeling.* Orlando, FL: Academic Press.

Heisenberg, W.Z. (1927). Quantum theory and measurement. *Physik*, 43, 172–198.

Johannsen, G., & Rouse, W.B. (1983). Studies of planning behavior of aircraft pilots in normal, abnormal, and emergency situations. *IEEE Transactions on Systems, Man and Cybernetics*, SMC-13 (3), 267–278.

Katzenbach, J.R., & Smith, D.K. (1993) *The Wisdom of Teams: Creating High-Performance Organizations.* Boston, MA: Harvard Business School Press.

Kuhn, T.S. (1962). *The Structure of Scientific Revolutions.* Chicago: University of Chicago Press.

Minsky, M. (1975). A framework for representing knowledge, in P.H. Winston, Ed., *The Psychology of Computer Vision.* New York: McGraw-Hill.

Nagel, K., Beckman, R.L., & Barrett, C.L. (1999). TRANSIMS for transportation planning. *Proceedings of the Sixth International Conference on Computers in Urban Planning and Urban Management*, Venice.

Newell, A., & Simon, H.A. (1972). *Human Problem Solving.* Englewood Cliffs, NJ: Prentice-Hall.

Pew, R.W., & Mavor, A.S. (Eds.). (1998). *Modeling Human and Organizational Behavior: Applications to Military Simulations.* Washington, DC: National Academy Press.

Rasmussen, J. (1983). Skills, rules and knowledge; signals, signs and symbols, and other distinctions in human performance models. *IEEE Transactions on Systems, Man & Cybernetics*, SMC-13 (3), 257–266.

Rasmussen, J. (1986). *Information Processing and Human-Machine Interaction.* New York: North-Holland.

Rasmussen, J., & Rouse, W.B. (Eds.). (1981). *Human Detection and Diagnosis of System Failures.* New York: Plenum Press.

Rasmussen, J., Pejtersen, A.M., & Goodstein, L.P. (1994). *Cognitive Systems Engineering.* New York: Wiley.

Rogers, S., Rogers, W., & Gregston, G. (1992). *Storm Center: The USS Vincennes and Iran Air Flight 655: A Personal Account of Tragedy and Terrorism.* Annapolis, MD: Naval Institute Press.

Rouse, W.B. (1977). A theory of human decision making in stochastic estimation tasks. *IEEE Transactions on Systems, Man, and Cybernetics*, SMC-7 (3), 274–283.

Rouse, W.B. (1980). *Systems Engineering Models of Human-Machine Interaction.* New York: North Holland.

Rouse, W.B. (1983). Models of human problem solving: Detection, diagnosis, and compensation for system failures. *Automatica*, 19 (6), 613–625.

Rouse, W.B. (2007). *People and Organizations: Explorations of Human-Centered Design.* New York: Wiley.

Rouse, W.B., & Hammer, J.M. (1991). Assessing the impact of modeling limits on intelligent systems. *IEEE Transactions on Systems, Man, and Cybernetics*, SMC-21 (6), 1549–1559.

Rouse, W.B., & Morris, N.M. (1986). On looking into the black box: Prospects and limits in the search for mental models. *Psychological Bulletin*, 100 (3), 349–363.

Rouse, W.B., & Rouse, R.K. (2004). Teamwork in the performing arts. *Proceedings of the IEEE*, 92 (4), 606–615.

Rouse, W.B., Hammer, J.M., & Lewis, C.M. (1989). On capturing humans' skills and knowledge: Algorithmic approaches to model identification. *IEEE Transactions on Systems, Man, and Cybernetics*, SMC-19 (3), 558–573.

Rouse, W.B., Cannon-Bowers, J.A., & Salas, E. (1992). The role of mental models in team performance in complex systems. *IEEE Transactions on Systems, Man, and Cybernetics*, 22 (6), 1296–1307.

Schank, R.C., & Abelson, R.P. (1977). *Scripts, Plans, Goals and Understanding.* Hillsdale, NJ: Lawrence Erlbaum.

Sheridan, T.B. (1992). *Telerobotics, Automation and Human Supervisory Control.* Cambridge, MA: MIT Press.

Sheridan, T.B., & Ferrell, W.R. (1974). *Man-Machine Systems: Information, Control, and Decision Models of Human Performance.* Cambridge, MA: MIT Press.

Simon, H.A. (1969). *The Sciences of the Artificial.* Cambridge, MA: MIT Press.

Swift, E. (2014). *Auto Biography: A Classic Car, an Outlaw Motorhead, and 57 Years of the American Dream.* New York: Harper Collins.

Zacharias, G.L., MacMillan, J., & Van Hemel, S.B. et al. (Eds.). (2008). *Behavioral Modeling and Simulation.* Washington, DC: National Academy Press.

Ziman, J. (1968). *Public Knowledge: The Social Dimension of Science.* Cambridge, UK: Cambridge University Press.

6

ECONOMIC PHENOMENA

INTRODUCTION

Table 6.1 lists the economic phenomena of interest for the six archetypal problems introduced in Chapter 2 and discussed in Chapter 3. All of these phenomena are concerned with demand, supply, prices, costs, and investments. This chapter addresses the state of knowledge in representing these phenomena.

The next two sections review concepts and models of microeconomics and macroeconomics, respectively. Within microeconomics, the theory of the firm and the theory of the market are reviewed. Two examples of microeconomic decision making are elaborated. The first focuses on optimal pricing to maximize profits. The second addresses the economics of investing in people.

Macroeconomic issues frame the context for microeconomic decisions. Gross domestic product (GDP) growth, tax rates, interest rates, and inflation can strongly affect the economic worth of alternative investments. For instance, inopportune combinations of these variables can significantly discourage the kinds of investments needed to address the six archetypal problems.

Modeling and Visualization of Complex Systems and Enterprises:
Explorations of Physical, Human, Economic, and Social Phenomena, First Edition. William B. Rouse.
© 2015 John Wiley & Sons, Inc. Published 2015 by John Wiley & Sons, Inc.

TABLE 6.1 Economic Phenomena Associated with Archetypal Problems

Problem	Phenomena	Category
Counterfeit parts	Macroeconomics of defense acquisition	Economic, macroeconomics
Financial system	Macroeconomics of demand and supply of investment products in general	Economic, macroeconomics
Healthcare delivery	Macroeconomics of demand, supply, and payment practices for healthcare	Economic, macroeconomics
Counterfeit parts	Microeconomics of supplier firms, including testing	Economic, microeconomics
Financial system	Microeconomics of firms creating specific investment products	Economic, microeconomics
Financial system	Microeconomics of risk and return characteristics of specific investment products	Economic, microeconomics
Financial system	Microeconomics of investor demand and prices of specific investment products	Economic, microeconomics
Healthcare delivery	Microeconomics of providers' investments in capacities to serve demand	Economic, microeconomics

The subsequent section considers behavioral economics. Traditional economists' assumptions about rational economic actors are often unrealistic. Research in behavioral economics has found the types of constraints that need to be added to models of human decision making. Unfortunately, addition of these constraints often renders the mathematics intractable. Simulation, rather than analytic solutions, can often be used to extend the range of usefulness of economic models. In situations where satisficing rather than optimizing is the norm, the usefulness of these models may be limited to providing insights rather than making valid predictions.

A final section addresses the economics of healthcare delivery. This example is laced with the microeconomic and macroeconomic phenomena discussed throughout this chapter. The microeconomics of the providers concerns fixed and variable costs as well as price sensitivities and economies of scale. The macroeconomics of large payers – specifically the federal government – concerns the transformation from a fee-for-service payment model to a pay-for-outcomes model. An example of how behavioral economics enters this picture is providers' great difficulties in escaping a "business as usual" mentality.

MICROECONOMICS

The six archetypal problems discussed throughout this book are laced with decisions by firms to invest in capacities to produce products or provide services, as well as the quality of these offerings. There are also decisions by individuals to respond to weather threats, avoid traffic congestion, and engage in chronic disease management programs. Some of these types of decisions were addressed in Chapter 5 on human phenomena. This chapter adopts a purely economic perspective on such decisions.

Microeconomics models are concerned with the behavior of firms and individuals in making decisions on the allocation of limited resources. These models are usually applied to markets where goods or services are bought and sold. Microeconomics examines how these decisions and behaviors affect the supply and demand for goods and services, which determines prices, and how prices, in turn, determine the quantity supplied and quantity demanded of goods and services. Microeconomic models of particular interest include the theory of the firm and theory of the market. This summary draws upon Sage and Rouse (2011).

Theory of the Firm

An enterprise's production function, f, is a specific mapping from or between the M input variables $\underline{X}$ to the production process and the output quantity produced, denoted by q.

$$q = f(\underline{X}) = f(x_1, x_2, \ldots, x_M) \tag{6.1}$$

If the firm prices these products or services at p per unit, then revenue is given by pq. If each input x_i has "wage" w_i, then profit Π is given by

$$\Pi(\underline{X}) = pf(\underline{X}) - \underline{W}^T \underline{X} - \text{fixed costs} \tag{6.2}$$

Realizing this profit depends on customers buying q units of products or services at price p. Whether or not this is a reasonable assumption depends on consumers' preferences, as well as alternative ways to spend their money.

Theory of the Market

Consumers consider multiple attributes, which we can denote by $y_1, y_2, \ldots, y_L$. Multiattribute utility theory considers the relationship between the set of attributes of an alternative, $\underline{Y}$, and the consumer's relative preferences for this alternative.

$$U(\underline{Y}) = U[u(y_1), u(y_2), \ldots u(y_L)] \tag{6.3}$$

$u(y_i)$ represents a consumer's utility function for attribute y_i. This function can take many forms, for example, from linear functions that imply each increment of y_i has the same change in utility, to diminishing returns functions where each increment of y_i yields decreasing changes of utility. The way in which all $u(y_i)$ are combined to yield $U(\underline{Y})$ can also take many forms, although linear (additive with weights) and multilinear forms are the most common (Keeney & Raiffa, 1976).

There may be K stakeholders whose preferences are of interest. This leads to the multiattribute, multistakeholder model.

$$U = U[U_1(\underline{Y}), U_2(\underline{Y}), \ldots U_K(\underline{Y})] \tag{6.4}$$

In this case, the attributes of the overall function U are the preferences of each stakeholder represented by $U_j(Y)$ for stakeholder j. Quite often, for the sake of simplicity and tractability, U is assumed to be a weighted sum of all the $U_j(Y)$.

Von Neumann and Morgenstern's (1944) axioms of rationality govern the formulation and combination of utility functions. This approach is prescriptive in that if one agrees with these axioms, then one *should* make decisions so as to maximize expected utility in the sense of the probabilistic notion of expected values. In contrast to this prescriptive approach is the descriptive approach of behavioral economics, which is reviewed later in this chapter.

Example – Optimal Pricing

Consider how these simple models can provide insights into investment issues. First, assume that the consumer's desires Q_D can be characterized by

$$Q_D(t) = Q_0[1 - \alpha P(t)] = \text{price sensitivity} \tag{6.5}$$

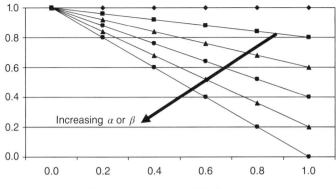

Figure 6.1 Q_D and VC Models

Note that this assumes that the consumer's utility function includes only one attribute – price. It further assumes that the consumer's utility function is a linear function of price. In reality, one would expect a nonlinear utility function, but over small price ranges, this is a reasonable approximation.

The profitability of the enterprise that provides these offerings, in quantities denoted by Q_P, can be characterized by

$$\pi(t) = [P(t)Q_D(t) - VC(t)Q_P(t)] - FC(t) = \text{profit} \tag{6.6}$$

where the variable and fixed costs can be modeled by

$$VC(t) = V_0[1 - \beta Q_P(t)] = \text{economy of scale} \tag{6.7}$$

$$FC(t) = [1 - GM(t)]P(t)Q_P(t) = \text{fixed costs} \tag{6.8}$$

$$GM(t) = \text{gross margin after G\&A, R\&D, etc.} \tag{6.9}$$

Note that equation (6.8) is very much an approximation of the fixed costs that are spread across all of production. This approximation makes the calculations that follow much more tractable.

The linear models for price sensitivity and economy of scales are shown on Figure 6.1. These linear approximations assume that variations of prices and quantities do not take on extreme values where nonlinearities would certainly emerge.

To find the optimal pricing, one takes the first partial derivative of profit with respect to price ($\partial \pi / \partial P$), sets this result equal to zero, and solves for the price. To determine whether this price is a maximum or minimum, one takes the second partial derivative with respect to price ($\partial^2 \pi / \partial P^2$), substitutes in

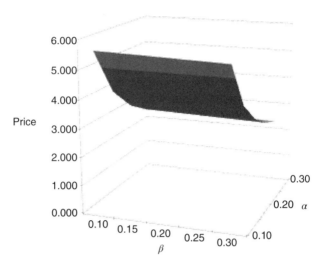

Figure 6.2   Price Surface

the optimal price, and examines the sign of the result. If the second derivative is negative, the price is a maximum; if it is positive, the price is a minimum.

Performing these operations, along with quite a bit of algebra, yields the price that optimizes profit (Rouse, 2010a).

$$P_{OPT} = (2\alpha\beta V_0 Q_0 - GM - \alpha V_0)/(2\,\alpha^2\beta V_0 Q_0 - 2\alpha\,GM) \qquad (6.10)$$

This represents the maximum profit when $\alpha\beta V_0 Q_0 \leq GM$, and the minimum loss when $\alpha\beta V_0 Q_0 > GM$. The latter condition holds when the parameters of the model are such that it is not possible to make a profit.

Figure 6.2 shows the sensitivity of the optimal price to market price sensitivities and economies of scale. Market price sensitivity, reflected in α, has a very strong effect while economy of scale, captured by β, has a much more modest effect. Figure 6.3 shows the sensitivity of profit, where the disparate effects of α and β are again evident. Thus, one can see that price and profits are highly affected by the price sensitivity of the market. In contrast, investments in improvements of economies of scale do not yield the same magnitudes of returns. Thus, a firm's discretionary monies may be better spent on convincing consumers, for example, via advertising, that they want the firm's product or service rather than investing these monies in efficiency.

Consider the firm's investment choices in terms of the model parameters. The firm could increase prices and profits by investing in decreasing α. They might do this by increasing advertising and/or addressing broader markets. This would tend to increase G&A (general and administrative) expenses and,

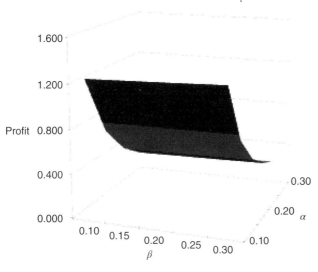

Figure 6.3 Profit Surface

hence, decrease GM (gross margin), which would, in turn, decrease profits. Thus, it may or may not be a good idea.

Another way to decrease α would be to invest in creating new proprietary products and services. The required R&D investments would have to be balanced against potential profits to determine whether these investments made sense. In general, R&D investments now yield increased revenues and profits in future years. If these future returns are steeply discounted, they may not be judged to be worthwhile.

The firm could invest in improving process efficiency and, consequently, increase β. As noted earlier, such investments do not tend to yield the magnitude of returns possible from decreasing α. However, in some markets, for example, military aircraft, advertising does not provide much leverage, the range of customers is constrained, and new proprietary offerings are rarely initiated without customer investments.

Another approach to increasing prices and improving profits is to increase GM. This can be accomplished by shrinking G&A, disinvesting in R&D, and minimizing other costs not directly related to current offerings. This can have an immediate and significant impact. However, it also tends to mortgage the future in the sense that the investments that are curtailed may be central to success in the future.

The application of this model to price controls in healthcare provided a variety of insights (Rouse, 2010a). Identification of the criteria for increased advertising, minimizing discretionary costs, or even market exit – as opposed to investing in increased efficiency – enabled predictions of these behaviors that were subsequently validated via compilations of studies where healthcare providers exhibited these exact behaviors.

Thus, in both this model and in reality, it is clear that there are important trade-offs among the possible ways to improve the firm's profitability. The ability to address and resolve these trade-offs is central to a firm's competitiveness and success. The ways in which decision makers frame and resolve such trade-offs are central to most of the six archetypal problems discussed in this book.

Example – Investing in People

A particularly interesting investment question concerns the extent to which it makes economic sense to invest in people – their training, education, health, safety, and work productivity (Rouse, 2010b). If the organizational entity making the investment is the same entity that receives the return on the investment, then quite often the investment makes economic sense. Examples include the safety of automobile production workers and the wellness of employees.

In contrast, if the investing entity does not receive the returns of such investments, this entity is likely to see such expenditures as costs rather than investments and, therefore, attempt to minimize such expenditures. The prevention and wellness program studied by Park and his colleagues (2012) provides an interesting example. They used computational models to redesign the delivery of this program to Emory University's employees. The redesigned program was projected to earn a 7% ROI (return on investment) due to decreased downstream employee healthcare costs and productivity losses.

The 7% ROI assumed that employees worked until age 65. Of course, healthcare costs do not end then, but Medicare becomes the payer. If it was assumed that people would live to age 80, the computational model projected a 30% ROI. The 23% difference benefitted Medicare rather than Emory. Consequently, Medicare should have been very interested in incentivizing Emory, as well as other employers, to invest in prevention and wellness programs. However, Medicare has no mandate to improve people's health before they are 65 and must enroll in Medicare.

More generally, there is reasonable agreement that we need a healthy, educated, and productive population that is competitive in the global marketplace. Yet, the numerous disconnects between who would have to invest and who would receive the benefits of these investments result in significant underinvestment. A broader perspective is needed to portray the critical connections that will convince key stakeholders of the viability of mechanisms whereby investments and returns can be aligned.

Summary

In this section, we have reviewed the theory of the firm and the theory of the market. Two examples were elaborated. One employed the theories of the firm and market to determine optimal pricing that maximizes profits. The second example focused on the economics of investing in people. We did not address competition, which is considered in Chapter 7.

MACROECONOMICS

Microeconomic decision making occurs in the environment of the current and expected macroeconomic context. Macroeconomics is concerned with the operation of the economy of a country or a region. The focus is on aggregate quantities such as the total amount of goods and services produced, total income earned, the level of employment of productive resources, and the level of prices. All of these constructs come together to yield GDP as shown in Figure 6.4.

There are several ways to calculate GDP. In theory, they should all yield the same answer, but measurement problems often result in variations of results. GDP, as shown in Figure 6.4, is the sum of consumption by households, investments in nonfinancial products, government spending excluding social transfers, and the net of exports minus imports.

Consumption by households is enabled by wages for labor, or transfers such as Social Security or unemployment benefits, that result in purchases of goods and services from producers. Both consumers and producers make investments in nonfinancial products such as houses or factories. Financial products are seen as simply swaps of one kind of currency for another. If these were included in GDP, the resulting metric could be arbitrarily inflated.

Government spends money to provide services such as roads, bridges, and security. It also spends money on social transfers that are counted once households spend these monies. Government receives tax payments from households (H), producers (P), capitalists (C), financiers (F), and landlords (L), as well as much smaller revenues from, for instance, entry fees at national parks.

Financiers provide money to capitalists in return for shares or equity. Capitalists enable producers to invest in capital goods in return for at least a share of the resulting profits. Landlords receive rents from households and producers. Obviously, all of these entities employ people and pay wages. These wages show up in GDP when households spend or invest these wages.

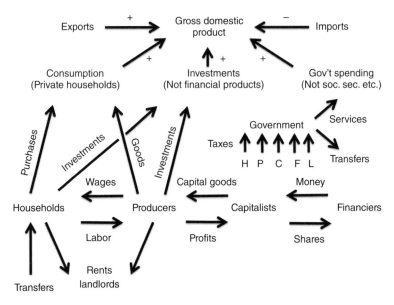

Figure 6.4   Elements of Gross Domestic Product (GDP)

As complicated as Figure 6.4 appears, it is very much a simplification. There are a variety of nuances that go into avoiding double or triple counting. Thus, for example, producers' goods and services are only counted when households purchase them. Similarly, as noted earlier, government transfers are only counted when households spend them.

Tax Rates, Interest Rates, and Inflation

Several important variables are not explicitly shown in Figure 6.4. In particular, tax rates, interest rates, and inflation do not appear. These parameters are particularly important to addressing many of the examples elaborated in this book. For example, the economic benefits of prevention and wellness programs or chronic disease management problems are highly sensitive to projected inflation because downstream savings due to keeping people healthy are greater when projected inflation is higher.

Edgeworth (1897) discusses what he terms the pure theory of taxation. Written before the 1913 establishment of US federal income tax, he focuses on taxes on the consumption of goods and commodities. Invoking the principle of equal sacrifice, he assesses impacts of taxes on each stakeholder. Equal sacrifice is defined as each taxpayer losing the same utility to the tax collector – the construct of utility was discussed earlier in the theory of the market in microeconomics.

Ramsey (1927) continues this line of reasoning. He is concerned with the following: given a bundle of goods and commodities, how to design the differing tax rates on each type of item to minimize loss of utility of those taxed. He mathematically shows that "the taxes should be such as to diminish the production of all commodities in the proportion" (p. 54). The key point here is that taxes on goods and commodities will diminish consumption of them.

Mirrlees (1971) explores the theory of optimal income taxation. He presents a mathematically rigorous approach to determining optimal progressive income taxes such that taxes impose equal utility losses on everyone. Individuals have a utility function for consumption (x) and time worked (y). An individual of ability n is paid ny for his or her work. The government imposes an income tax percentage of $c(ny)$ on this income. The overall result is that a $c(ny)$ function that is linear in rate progression is near optimal, unless "the supply of highly skilled labor is inelastic."

Mirrlees suggests, however, that greater equality could be achieved by taxing people by ability to earn income, perhaps proportional to IQ, rather than actual income as "an effective method for offsetting the unmerited favors that some of us receive from our genes and family advantages." The difficulty that Mirrlees is trying to overcome with this idea is that high tax rates are likely to discourage high-ability people from maximizing their earnings, thereby undercutting the tax revenues that were expected from them. Taxing by IQ, in theory at least, implies that people would have to pay taxes on what they could earn even if they chose not to earn these amounts. It is, of course, very difficult to imagine this ever being implemented.

Feldstein (1982) and Summers (1981) address the impacts of tax rates and inflation on investments. These pieces were written during a period of high inflation in the US economy. Feldstein argues that "Inflation creates fictitious income for the government to tax" (p. 154). Taxes are paid on inflated returns while investors gain real returns. In this way, a positive real return can become a net loss once taxes are deducted. This is true for both investments by businesses and savings by individuals.

The real interest rate equals the nominal interest rate minus inflation. High inflation rates can lead to negative real interest rates. This encourages home buying, as well as purchase of consumer durables, to gain both the tax deduction for *nominal* interest paid and increased property values due to inflation. These motivations further drive up home prices. Note that this makes sense if homeowners were not expecting to pay capital gains taxes on the inflated home value. Overall, inflation and tax policy had discouraged business investment and encouraged residential real estate investment.

Summers (1981) discusses trends in business investments versus those made to satisfy regulations, for example, pollution control. The investment rate in the 1975–1979 period of 3% was the lowest in three decades. While total taxes (on corporate profits, dividends, and capital returns) declined from 71.5% in 1953 to 52.7% in 1979, inflation soared. "In a tax-less world, firms invest as long as each dollar spent purchasing capital raises the market value of the firm by more than one dollar" (p. 77). However, as Feldstein notes, high inflation rates undermine this possibility. Summers concludes that "The most desirable investment incentives are those that operate by reducing the effective purchase price of new capital goods" (p. 118).

Much more recently, Mankiw et al. (2009) discuss optimal taxation in theory and practice. They argue that "The social planner has to make sure the tax system provides sufficient incentive for high-ability taxpayers to keep producing at the high levels that correspond to their ability, even though the social planner would like to target this group with higher taxes."

They report eight lessons from their review:

- Optimal marginal tax rate schedules depend on the distribution of ability.
- The optimal marginal tax schedule could decline at high incomes.
- A flat tax, with a universal lump-sum transfer, could be close to optimal.
- The optimal extent of redistribution rises with wage inequality.
- Optimal taxes should depend on personal characteristics as well as income.
- Only final goods ought to be taxed, and typically they ought to be taxed uniformly – this is the essence of a value-added tax.
- Capital income ought to be untaxed, at least in expectation.
- In stochastic, dynamic economies, optimal tax policy requires increased sophistication.

Anzuini et al. (2010) consider the impact of monetary policy on commodity prices. They report that "Global monetary conditions have often been cited as a driving factor of commodity prices. They investigate the empirical relationship between US monetary policy (i.e., low interest rates) and commodity prices by means of a standard auto-regression methodology, commonly used in analyzing the effects of monetary policy shocks. The results suggest that expansionary US monetary policy shocks drove up the broad commodity price index and all of its components. While these effects were significant, they, however, did not appear to be overwhelmingly large. This finding is also confirmed under different identification strategies for the monetary policy shock."

Ferede and Dahlby (2012) examined the impact of the Canadian provincial governments' tax rates on economic growth using panel data covering the period 1977–2006. They found that a higher provincial statutory corporate income tax (CIT) rate was associated with lower private investment and slower economic growth. Their empirical estimates suggest that a 1% point cut in the corporate tax rate is related to a 0.1–0.2% point increase in the *annual* growth rate. Their results also indicate that switching from a retail sales tax to a "harmonized" sales tax that is harmonized with the federal value-added sales tax boosts provincial investment and growth.

They used regression models and panel data. Their methods discussion shows that there is some subtlety to support these seemingly intuitively results. They used simulation to predict long-term impact, which would play out over decades. The results indicate that in the long run, British Columbia's per capita GDP with the CIT tax cut will be about 16% higher than in the absence of the tax cut.

James (2009) considers investment incentives. He indicates that "Governments make extensive use of investment incentives in an effort to attract investments. Their effectiveness has been the subject of intense debate, and little consensus has emerged. These disparate views are not surprising given that tax and nontax incentives are just one of the many factors that influence the success of investments. Countries typically pursue growth-related reforms using a combination of approaches, including macroeconomic policies, investment climate improvements, and industrial policy changes – including investment incentives. If such reforms have led to growth, it is difficult to attribute it solely to incentives."

"Every investment incentive policy has potential costs and benefits. The benefits arise from higher revenue from possibly increased investment, and social benefits – such as jobs, positive externalities, and signaling effects – from this increased investment. The costs are due to revenue losses from investments that would have been made even without the incentives, and indirect costs such as economic distortions and administrative and leakage costs."

James reaches the following conclusions about investment incentives:

- On their own, such incentives have limited effects on investments. Countries must also dedicate themselves to improving their investment climates.
- If used, investment incentives should be used minimally – mainly to address market failures and generate multiplier effects.
- Incentives should be awarded with as little discretion and as much transparency as possible, using automatic legal criteria.
- To the extent possible, incentives should be linked to investment growth (i.e., based on performance), and tax holidays should be avoided.

Carroll and Prante (2012) address the impact of the increase in tax rates for high-income taxpayers resulting in part due to the sunset of elements of the 2001 and 2003 tax cuts, as well as the increase in the Medicare tax and its expansion to unearned income for high-income earners. The concern over the top individual tax rates was a focus, in part, because of the prominent role played by flow-through businesses – S corporations, partnerships, limited liability companies, and sole proprietorships – in the US economy and the large fraction of flow-through income that is subject to the top two individual income tax rates. These businesses employ 54% of the private sector work force and pay 44% of federal business income taxes. The number of workers employed by large flow-through businesses is also significant: more than 20 million workers are employed by flow-through businesses with more than 100 employees.

This analysis employed the Ernst & Young General Equilibrium Model of the US Economy to examine the impact of the increase in the top tax rates in the long run. This report examined four sets of provisions that will increase the top tax rates:

- The increase in the top two tax rates from 33% to 36% and 35% to 39.6%.
- The reinstatement of the limitation on itemized deductions for high-income taxpayers (the "Pease" provision).
- The taxation of dividends as ordinary income and at a top income tax rate of 39.6% and increase in the top tax rate applied to capital gains to 20%.
- The increase in the 2.9% Medicare tax to 3.8% for high-income tax-payers and the application of the new 3.8% tax on investment income including flow-through business income, interest, dividends, and capital gains.

With the combination of these tax changes at the beginning of 2013, the top tax rate on ordinary income would rise from 35% in 2012 to 40.9%, the top tax rate on dividends would rise from 15% to 44.7%, and the top tax rate on capital gains will rise from 15% to 24.7%. These higher tax rates would result in a significant increase in the average marginal tax rates (AMTR) on business, wage, and investment income, as well as the marginal effective tax rate (METR) on new business investment. This report finds that the AMTR increases significantly for wages (5.0%), flow-through business income (6.4%), interest (16.5%), dividends (157.1%), and capital gains (39.3%). The METR on new business investment increases by 15.8% for the corporate sector and 15.6% for flow-through businesses.

They predict that these higher marginal tax rates will result in a smaller economy, fewer jobs, less investment, and lower wages. Specifically, this report finds that the higher tax rates will have significant adverse economic effects in the long run: lowering output, employment, investment, the capital stock, and real after-tax wages when the resulting revenue is used to finance additional government spending.

Through lower after-tax rewards work, the higher tax rates on wages reduce work effort and labor force participation. The higher tax rates on capital gains and dividend increase the cost of equity capital, which discourages savings and reduces investment. Capital investment falls, which reduces labor productivity and means lower output and living standards in the long run.

- Output in the long run would fall by 1.3%, or $200 billion, in today's economy.
- Employment in the long run would fall by 0.5% or, roughly 710,000 fewer jobs, in today's economy.
- Capital stock and investment in the long run would fall by 1.4% and 2.4%, respectively.
- Real after-tax wages would fall by 1.8%, reflecting a decline in workers' living standards relative to what would have occurred otherwise.

These results suggest real long-run economic consequences for allowing the top two ordinary tax rates and investment tax rates to rise in 2013. This policy path can be expected to reduce long-run output, investment, and net worth.

Thomas Piketty has focused on income inequality and capital taxation (Piketty & Saez, 2012; Piketty, 2014), with his 2014 book receiving an enormous amount of attention. They focus on "socially-optimal capital taxation," on both savings and bequests, to deal with the problem of a "large concentration of inherited capital ownership." They assert that "Inequality permanently arises from two dimensions: differences in labor income due to differences in ability, and differences in inheritance due to differences in parental tastes for bequests and parental resources."

They use their formulation of this optimization problem to show that the "socially optimal" tax rate on inheritances, TB, can range from 40–60% to 70–80% when bequests are highly likely. Increasing TB allows decreasing the tax rate for labor, TL. For a 20% bequest probability, TB = 73% and TL = 22%. For a 5% bequest probability, TB = 18% and TL = 42%. The probability of a bequest has been found to be negatively correlated with economic growth. Note that results depend on social weight given to those receiving zero bequests – in all countries for which they have data, the bottom 50%

of people (economically) have about 5% of the inherited capital. Their model also includes a tax rate for capital gains, TK. They consider how people shift income from monies subject to TL versus TK. They find that the optimal TK increases with uncertainty about future returns due to TB. They also consider a consumption tax, TC. This tax can enable TL < 0, which implies a labor subsidy for low-income people.

Of course, whatever schemes are proposed for TB, TL, TK, and TC, high-ability people will figure out how to take advantage of these schemes, effectively thwarting income redistribution. People with high incomes and/or wealth will have the resources to hire high-ability people to figure this out for them. Thus, any scheme is subject to "gaming" and adaptation will inevitably be necessary.

Macroeconomic Models

There is a class of macroeconomic models intended to address the dynamic nature of the economy, as well as the stochastic or probabilistic nature of the forces driving the dynamics. This section discusses a sampling of such models.

Amisano and Geweke (2013) compare models for predicting economic growth. Three models are compared:

- Dynamic factor model – autoregressive time series model of factors
- Dynamic stochastic general equilibrium model – version of Monte Carlo
- Multivariate regression model

They fit these models to data and then use the models to predict means and standard deviation of the following macroeconomic variables:

- Consumption growth
- Investment growth
- Income growth
- Hours worked index
- Inflation growth
- Wage growth
- Federal funds rate.

Data were from the period 1951–2011. Predictions were for one-quarter ahead over the period 1966–2011. No one model was consistently superior. The best comparisons came from a weighted average of the predictions of all three models.

Gali and Gertler (2007a, 2007b) consider the impacts of monetary policy, which is defined as follows: "Monetary policy is the process by which the monetary authority of a country controls the supply of money, often targeting a rate of interest for the purpose of promoting economic growth and stability."

Considering approaches to modeling, they conclude that "Econometric models fit largely on statistical relationships from a previous era did not survive the structural changes of the 1970s. One could have little confidence that the parameter estimates would be stable across regimes."

They argue for two key implications of new modeling frameworks:

- "Monetary transmission depends critically on private sector expectations of the future path of the central bank's policy instrument, the short-term interest rate."
- "The natural (flexible price equilibrium) values of both output and real interest rate provide important reference points for monetary policy – and may fluctuate considerably."

Noting that, "An important challenge for central banks is tracking the natural equilibrium of the economy, which is not directly observable," they present their stochastic dynamic general equilibrium model of aggregate demand, aggregate supply, and policy. In this model, firms decide on prices and how much to produce. Expectations depend on both current and anticipated policies. "Each period the central bank chooses a target for the short-term interest rate as a function of economic conditions. To attain that rate, the central bank adjusts the money supply to meet the quantity of money demanded at the target interest rate."

They observe that "Monetary policy, i.e., interest rates, is central to the model, but the money supply only serves to provide a unit of account." They advocate use of their model for projecting the natural equilibrium of the economy.

Brunnermeier and Sannikov (2013) address the stability of the financial sector. They define a crisis regime as when market volatility, credit spreads, and financing activity change drastically. They argue that "Temporary shocks can have persistent effects on economic activity as they affect the net worth of levered agents, and financial constraints. Net worth takes time to rebuild. Financial frictions lead to amplifications of shocks, directly through leverage and indirectly through prices. Small shocks can have potentially large effects on the economy. The amplification through prices works through adverse feedback loops, as declining net worth of levered agents leads to drops in prices of assets concentrated in their hands, further lowering these agents' net worth."

Their model focuses on length, severity, and frequency of crises. They represent the dynamics of the market as a differential equation or, equivalently, a difference equation as shown in equation (6.11). The driving force is Brownian shocks.

$$\Delta C = (G - D) \cdot C \cdot \Delta t + \sigma \cdot C \cdot \Delta Z \tag{6.11}$$

where C = capital, G = growth, D = depreciation, σ = volatility, and Z represents exogenous Brownian shocks. To solve for equilibrium, they first employ agent utility maximization and market clearing conditions and then derive the equilibrium dynamics in terms of experts' wealth shares.

Implications of their model of the system dynamics include:

- The system's reaction to shocks is highly nonlinear – resilient for shocks near steady state, but unusually large shocks are strongly amplified.
- The system's reactions to shocks is asymmetric – positive shocks at steady state lead to larger payouts and little amplification while large negative shocks are amplified into crises.
- Endogenous risk, that is, risk self-generated by the system, dominates the volatility and affects experts' precautionary moves.
- The extent and length of slumps is stochastic and significantly increases the amplification and persistence of adverse shocks.
- After moving through a high volatility region, the system can get trapped in a recession with low growth and misapplication of resources.

Summary

Macroeconomic issues are important to the types of problems discussed in this book because they frame the context for microeconomic decisions. GDP growth, tax rates, interest rates, and inflation can strongly affect the economic worth of alternative investments. For instance, inopportune combinations of these variables can significantly discourage the kinds of investments needed to address the six archetypal problems. In general, macroeconomics provides the context within which each of the archetypal problems needs to be addressed.

BEHAVIORAL ECONOMICS

Behavioral economics addresses how real people make decisions to pursue a particular path, allocate resources, and so on. The ideal "economic man" is replaced with the reality of biases, heuristics, satisficing, and so forth. This is most often studied at the microeconomic level of individual choice and personal preferences. However, those responsible for macroeconomic policy decisions are, of course, subject to the same human inclinations.

Behavioral finance is a subdiscipline within behavioral economics. This subdiscipline is concerned with explaining why participants in financial markets systematically behave in ways that conflict with assumptions of rational market theorists. Such behaviors create market inefficiencies that other participants exploit – termed arbitrage.

These aberrant behaviors – relative to rational market assumptions – include under- or overreactions to information as causes of market trends and, in extreme cases, subsequent bubbles and crashes. Such reactions have been attributed to attention limitations, overconfidence, overoptimism, and herding. In general, these types of phenomena enable the behaviors seen in the archetypal problem of the financial system discussed earlier.

Other key observations include the asymmetry between decisions to acquire, or keep resources, and loss aversion, the unwillingness to liquidate a valued investment. Loss aversion results in investor reluctance to sell equities that would result in losses. For example, homeowners may be unwilling to sell houses at depressed prices below what they paid for the house.

Given the nature of the archetypal problems discussed in this book, the behavioral economic phenomena of most interest involve individual decision making. In particular, the concern is with decision-making behaviors that conflict with the standard "rational" assumptions associated with microeconomics and macroeconomics. It is useful to begin by considering a range of perspectives on rationality (Deising, 1962).

Economic rationality is premised on maximization of economic value. Complementary to this perspective is technical rationality that focuses on maximizing goal achievement. There is often a confluence of financial and engineering agendas in developing complex public–private systems such as defense systems and urban infrastructures.

However, other forms of rationality often complicate things. Social rationality is concerned with organizational preservation and betterment. Political rationality emphasizes preservation of existing balances of interests, with incremental changes. Legal rationality stresses compliance with rules that are complex, consistent, precise, and detailed. These three forms of rationality often can compromise what otherwise would seem like attractive financial–technical trade-offs.

There is a rather broad range of models of human decision making. Hammond and colleagues (1980) have compiled and contrasted six different approaches. Three approaches originated in economics – decision theory (DT), behavioral decision theory (BDT), and psychological decision theory

TABLE 6.2   Theories of Human Decision Making

Model	Proponents	Goal	Use
DT	Keeney and Raiffa	Achievement of logical consistency	Aid decision makers
BDT	Edwards	Describe and explain departures from optimality	Aid decision makers to avoid departures from optimality
PDT	Kahneman and Tversky	Provide psychological explanations of behaviors	Aid decision makers via "debiasing"
SJT	Hammond	Describe human judgment in ambiguous social situations	Aid decision makers to understand relationship between judgment and cues – "policy capturing"
IIT	Anderson	Cognitive algebra to explain stimulus–response behaviors in social settings	Show that algebra fits!
AT	Jones and Kelley	Describe how humans explain behavior of objects or events	Show how humans attribute causes

(PDT). The other three approaches were founded in psychology – social judgment theory (SJT), information integration theory (IIT), and attribution theory (AT). Table 6.2 contrasts these six theories of human decision making.

As discussed earlier in this chapter, DT is prescriptive – it helps one to be logically consistent and compliant with the axioms of rationality. The other five theories are descriptive, focused on actual and typically unaided human decision making. The one exception is the use of the theories of psychological decision making to prescribe how to "debias" human decisions.

The descriptive theories were motivated by desires to understand how people actually make decisions because they often depart from the prescriptions of DT. Edwards (1954) and later Tversky and Kahneman (1974, 1979) led the efforts to understand subjective probabilities and heuristics, both of which can lead to biases. Hammond and colleagues' (1980) policy capturing mapped decisions to the cues upon which they were based to provide decision makers estimates of the weightings they were using. Anderson's (1981) cognitive algebra posited a set of mathematical operations between cues and decisions. Finally, Jones and Harris (1967) and Kelley and Michela (1980) explored how humans attribute cause to observed events.

Prospect Theory

Kahneman and Tversky conducted a range of investigations into the heuristics and biases humans employ when making decisions (Tversky & Kahneman, 1974; Kahneman et al., 1982; Kahneman, 2011).

"Representativeness" is a bias due to insensitivity to the a priori probabilities of outcomes. For example, if asked about the likelihood of an airplane crash versus an automobile crash on any given day, one would tend to overestimate the likelihood of a plane crash and underestimate the likelihood of a car crash unless one clearly understood the a priori probabilities of such events.

"Availability" is a bias due to the retrievability of instances. If one can more easily imagine what an astronaut would look like than a vacuum cleaner salesman, then one would tend to overestimate the likelihood of encountering an astronaut and underestimate the likelihood of a vacuum cleaner salesman.

"Anchoring and Adjustment" is a bias affected by initial framing of a question, perhaps in terms of initial estimates of a price or probability. For example, the initial price of an object, say a home or a car, will cause people to perceive that they are getting a good deal if the final price is significantly lower than the initial price despite the strong possibility that the initial price was substantially inflated.

"Motivational Bias" is the tendency for estimates of the likelihood of outcomes to be affected by the desirability of outcomes. Optimists tend to overestimate the likelihood of desired outcomes and underestimate the likelihood of undesirable outcomes. Pessimists tend to underestimate the likelihood of desired outcomes and overestimate the likelihood of undesirable outcomes.

Kahneman and Tversky (1979) later introduced prospect theory to explain various divergences of decision making from classical economic theory. Prospect theory includes an editing stage and an evaluation stage. Risky situations are simplified using the types of heuristics just discussed in the editing stage. Risky alternatives are evaluated in the evaluation phase using the following principles:

- *Reference dependence*. Outcomes are compared to a desired reference point and classified as "gains" if greater than the reference point and "losses" if less than the reference point.

- *Loss aversion.* Losses hurt more than equivalent gains. Kahneman and Tversky found that losses hurt more than twice as much as equivalent gains.
- *Probability weighting.* Decision makers overweight small probabilities and underweight large probabilities, resulting in an inverse-S-shaped weighting function.
- *Diminishing sensitivity.* As the magnitudes of gains and losses relative to the reference point increase in absolute value, the incremental effect on the decision maker's utility decreases.

Prospect theory explains the same phenomena explained by traditional DT. However, prospect theory has also been used to explain a range of phenomena that DT has difficulty explaining because they appear to violate the axioms of rationality. It can be argued that humans are constrained optimal, as elaborated in Chapter 5, subject to the constraints of the heuristics and biases discussed earlier and the elements of prospect theory just elaborated. Thus, their rationality is bounded (Simon, 1972).

Tversky and Kahneman (1992) extended prospect theory, now calling it cumulative prospect theory. They eliminated the editing phase and focused just on the evaluation phase. The main feature was the application of the nonlinear probability weighting function to cumulative probabilities. Other additions to this theory, over time, have included phenomena such as over-confidence and limited attention.

Risk Perception

Slovic et al. (1982) focused on quantifying and predicting humans' perceptions of risk. Their findings complement those of Kahneman and Tversky. For example, the greater people perceive a benefit, the greater the tolerance for a risk. For products from which people derive pleasure, they tend to judge its benefits as high and its risks as low. For activities that are disliked, the judgments are opposite. Risk perception is highly dependent on intuition, experiential thinking, and emotions. They found that three factors influenced perceptions of risk: the degree to which a risk is understood, the degree to which this risk evokes a feeling of dread, and the number of people exposed to the risk. Perceived risk is related to the extent to which these factors are present.

The construct of risk has long been considered to include two components, the probability of consequences and the magnitude of consequences

(Kaplan & Garrick, 1981). Some pundits take this further and define risk as follows.

$$\text{Risk} = \text{Probability of consequences} \times \text{magnitude of consequences} \quad (6.12)$$

This leads to the interesting question of whether people seemingly taking inordinate risks do so because they underestimate the probability or magnitude of consequences – or both.

These phenomena are very much relevant to the urban resilience problem archetype discussed in several places in this book. Blumberg, as discussed in Chapter 4, very accurately predicted the water levels that would result from Hurricane Sandy – 13 feet above normal, a 300-year high. Many people either did not understand the implications of these projections or dismissed them. Why?

Given the frequent experiences of flooding, it would be reasonable to assume that most people understood the consequences. Recovering from flooded basements is a common experience in greater New York City – the city plus northeast New Jersey and southwest Connecticut. Thus, it is more likely that they discounted the probability of the consequences rather than the magnitude of the consequences.

In fact, as the storm approached and evacuation warnings were given, it was common to hear, "You told us the same things about Hurricane Irene last year, and not much of anything happened." Indeed, the national television news channels had "cried wolf" repeatedly as Irene approached. Thus, for Hurricane Sandy, people were clearly discounting the probability of consequences.

Attribution Errors

AT is concerned with how humans attribute causes to observed events (Jones & Harris, 1967). Not surprisingly, people often make attribution errors. Determining causation can be quite difficult. Of particular interest is the phenomenon called "fundamental attribution error."

It is easy to explain this construct with an instance of this type of error that has been empirically validated repeatedly. When you are successful, there is a strong tendency for you to attribute this success to your great skill and sustained effort. When you fail, you will tend to attribute such failure to bad luck.

In contrast, when you observe the successful accomplishments of others, you will have a tendency to attribute success to good luck. Failures by others, you will tend to attribute to their lack of sufficient skill and adequate effort. This asymmetry is striking. It will be revisited in Chapter 7.

Gladwell (2008) in his popular book *Outliers* asserts that there are three determinants of success – ability, motivation, and circumstances. Success is due to superior abilities, motivation to work hard, and being in the right place at the right time.

Bill Gates of Microsoft and Steve Jobs of Apple were talented microcomputer enthusiasts at just the right time in that industry's development. Great success can be achieved by taking advantage of the times, along with having the abilities and motivation to exploit the opportunities available. Gladwell discusses numerous examples of this phenomenon.

Management Decision Making

What do the aforementioned departures from traditional axioms of rationality imply for how managers make decisions for their organizations? A wide variety of organizational decisions are laced throughout the six archetypal problems discussed throughout this book. How well do decision makers plan, organize, coordinate, and control the activities of their organizations?

Mintzberg (1975) published a classic analysis of the jobs of managers. Based on a variety of studies, he concluded that planning, organizing, and so on is not the right characterization of managers' jobs. He instead described their jobs in terms of three roles – interpersonal, informational, and decisional. He further subdivided these roles:

- Interpersonal – Figurehead, Leader, Liaison
- Informational – Monitor, Disseminator, Spokesperson
- Decisional – Entrepreneur, Disturbance Handler, Resource Allocator, Negotiator.

This characterization shows how managers' jobs differ substantially from being primarily dispassionate analytical optimizers of economic outcomes. Only in one of ten roles – resource allocator – does the notion of optimization seem relevant.

Simon (1972) considers this possibility in his exploration of humans' bounded rationality. He argues that human decision making is limited by information available, cognitive limitations, and the time available. Consequently, decision makers "satisfice" rather than optimize. Thus, for example, decision makers allocate resources in a manner that is satisfactory in that it makes sense and is acceptable.

Bounded rationality can be augmented by decision support that brings together information, performs calculations impossible for unaided humans, and does all this within the available time for making a decision. When this is feasible, an economically optimal solution may be feasible. This works for routing delivery trucks, for example. However, optimization is only possible when objectives are very clear, the dynamics of the system are understood, and constraints can be formalized. This is seldom possible for most realistically complex problems.

Human Intuition

The foregoing would lead one to think that humans are rather limited, at least relative to what economists would like to assume in their microeconomic or macroeconomic models. Fortunately, Klein's (2003) pioneering studies in a variety of complex domains have shown that humans can be excellent pattern recognizers despite often being inadequate optimizers.

Figure 6.5 is a slightly modified version of Klein's depiction (Klein, 2003, p. 16). Situations generate cues that enable humans to recognize patterns that activate scripts for responding to the situation. Humans mentally simulate these scripts using their mental models as elaborated in Chapter 5. If the outcomes of these simulations are satisfactory, humans act and the situation evolves to generate updated cues, and so on.

Klein was particularly interested in the ways in which intuitions guide the actions of experts. For example, fire commanders at the scene of a fire often have to decide and act quickly regarding whether or not to enter a burning building. Klein found that experts' intuition could be relied upon in such situations. He reached the same conclusion in several other domains.

If we think about human decision making broadly, Klein's conclusions about intuition have to be correct, at least much of the time. Consider the aforementioned 10 elements of Mintzberg's characterization of management decision making. Without recognition-primed decision making, managers would not be able to make it through their workday. If every task had

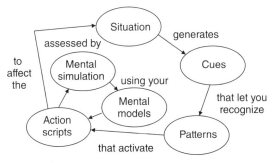

Figure 6.5 Klein's Recognition-Primed Decision-Making Model (*Source*: Reproduced with permission from Klein (2003). Copyright © 2003, Doubleday.)

to be approached analytically, life for mangers and everyone would be overwhelming.

Note that Gladwell discusses Klein's findings in depth in *Blink* (2005). Gladwell's easily readable and often compelling treatises provide an interesting starting point for those interested in behavioral economics. In particular, his *Tipping Point* (2000), *Blink* (2005), and *Outliers* (2008) provide a quick appreciation of several concepts discussed in this section.

Intuition versus Analysis

In light of the aforementioned factors, when should one rely on intuition and when should one revert to more analytical approaches such as described earlier in this chapter? I developed the simple framework in Table 6.3 for addressing this question (Rouse, 2007). The answer depends on the situation of concern.

If the situation is familiar and one has addressed it frequently, intuition should work very well. Such situations dominate much of everyday life. Analysis is not needed to get dressed in the morning, choose breakfast, drive to work, and handle the many routine tasks encountered every day.

For situations that are familiar but seldom encountered, such an annual preparation of income tax returns, off-the-shelf tools can be employed, leavened with intuition to assure that results make sense. If, for instance, the calculated tax due is greater than overall income, something is wrong somewhere.

For unfamiliar situations that will be encountered frequently, analysis may be needed initially to, for example, find the best route to work in a new city. With good feedback, intuition will develop and analysis will no longer be needed. Without feedback, however, one will not know what really works and intuition will be stymied.

Unfamiliar situations that are rarely encountered often warrant an analytical approach, particularly if the consequences are of great import. Buying a house, moving to a new city, and deciding where to retire are good examples.

TABLE 6.3   Intuition versus Analysis for Different Situations

Situation	Familiar	Unfamiliar
Frequent	Intuition should work very well	Intuition will develop with good feedback
Infrequent	Intuition and off-the-shelf analysis	Analysis with intuition for sanity checks

Similarly, buying a competitor, opening a new factory, and selling the company are excellent examples of where analytical methods are valuable. Of course, intuition is still important for "sanity checks." Senior executives seldom make major decisions just because of the bottom lines on spreadsheets.

Summary

Traditional economists' assumptions about rational economic actors are often unrealistic. Research in behavioral economics has found the types of constraints that need to be added to models of human decision making. Unfortunately, addition of these constraints often renders the mathematics intractable. Simulation, rather than analytic solutions, can often be used to extend the range of usefulness of economic models. In situations where satisficing rather than optimizing is the norm, the usefulness of these models may be limited to providing insights rather than making valid predictions. Fortunately, as illustrated in Chapters 8 and 9, insights rather than predictions may often be the key to problem solving and decision making.

ECONOMICS OF HEALTHCARE DELIVERY

The last four decades have seen enormous increases in healthcare costs. Specifically, *real* healthcare costs have tripled as a percent of the GDP in the period 1965–2005, with half of this growth due to technological innovation (CBO, 2008). The magnitude of these increases has led some pundits to conclude that the nation is "running on empty" (Peterson, 2005). There seems to be virtually unanimous agreement that something has to change significantly.

Figure 6.6 summarizes the overall phenomena discussed in the CBO report. Technological inventions become market innovations as they increase in effectiveness and associated risks decrease. This results in increased use, which leads to increased expenditures. In parallel, increased efficiency via production learning (more on this next) leads to decreased cost per use, although not enough to keep up with growing use rates in healthcare. Finally, increased use yields improved care that leads to longer lives and increased chances of again employing the technology of interest.

The concern in this illustrative example is how to control the phenomena depicted in Figure 6.6. More specifically, what efficiencies must be realized to steadily decrease the cost per use to offset the constantly increased number of uses and thereby enable affordable healthcare? Equation (6.13) provides the basis for addressing this question from a broader perspective than just efficiency.

$$\text{Total cost} = \text{cost per use} \times \text{number of uses} \qquad (6.13)$$

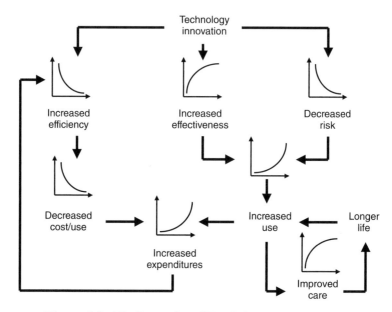

Figure 6.6 The Dynamics of Escalating Healthcare Costs

We can decreases overall healthcare costs by decreasing the cost per use and/or decreasing the number of uses. Decreasing cost per use involves a quest for efficiency. In most industries, the costs of technology-based offerings decrease due to production learning. Every doubling of the number of units produced leads to unit costs decreasing to 70–90% of what they had been. Subsequent doublings lead to further decrease in unit costs.

This has not worked in healthcare for two reasons. First, the fee-for-service payment model strongly incentivizes delivering more services and certainly does not incentivize price decreases. Second, any efficiencies due to production learning are overwhelmed by labor inefficiencies. The key to decreasing cost per use is pursuit of learning curves that decrease the cost of labor in healthcare delivery and there is a wealth of ways this could be accomplished (Rouse, 2009).

However, steadily decreasing the costs per use, no matter how challenging, is unlikely to compensate for exponentially increasing number of uses. To reign in such growth, "population health" initiatives such as prevention and wellness programs and chronic disease management programs are needed (Park et al., 2012; Rouse & Serban, 2014). Such programs have been found to substantially decrease visits to emergency departments and in-patient admissions to hospitals.

This has resulted in decreasing the number of uses of healthcare providers' capacities. In fact, many hospitals are seeing declining revenues as utilization of capacities decreases. To compensate for this trend, providers are seeking to serve larger populations of patients via alliances with or acquisitions of other providers. This process is expected to lead to many fewer, but substantially larger integrated health systems (Rouse & Cortese, 2010).

This example is laced with the microeconomic and macroeconomic phenomena discussed throughout this chapter. The microeconomics of the providers concern fixed and variable costs as well as price sensitivities and economies of scale. The macroeconomics of large payers – specifically the federal government – concern the transformation from a fee-for-service payment model to a pay-for-outcomes model. An example of how behavioral economics enters this picture is providers' great difficulties in escaping a "business as usual" mentality.

CONCLUSIONS

The chapter has reviewed concepts and models of microeconomics and macroeconomics. Within microeconomics, the theory of the firm and the theory of the market were reviewed. Two examples of microeconomic decision making were elaborated. The first focused on optimal pricing to maximize profits. The second addressed the economics of investing in people.

Macroeconomic issues frame the context for microeconomic decisions. GDP growth, tax rates, interest rates, and inflation can strongly affect the economic worth of alternative investments. For instance, inopportune combinations of these variables can significantly discourage the kinds of investments needed to address the six archetypal problems discussed in this book.

We next considered behavioral economics. Traditional economists' assumptions about rational economic actors are often unrealistic. Research in behavioral economics has found the types of constraints that need to be added to models of human decision making. Unfortunately, addition of these constraints often renders the mathematics intractable. Simulation, rather than analytic solutions, can often be used to extend the range of usefulness of economic models. In situations where satisficing rather than optimizing is the norm, the usefulness of these models may be limited to providing insights rather than making valid predictions.

Finally, the economics of healthcare delivery was discussed. This example was laced with the microeconomic and macroeconomic phenomena discussed throughout this chapter. The microeconomics of the providers concern fixed and variable costs as well as price sensitivities and economies of scale. The macroeconomics of large payers – specifically the federal government – concern the transformation from a fee-for-service payment model to a pay-for-outcomes model. Behavioral economics is relevant here in terms of how providers address the transformation of their enterprises.

REFERENCES

Amisano, G., & Geweke, J. (2013). *Prediction Using Several Macroeconomic Models*. Brussels: European Central Bank, Working Paper No. 1537.

Anderson, N. H. (1981). *Foundations of Information Integration Theory*. New York: Academic Press.

Anzuini, A., Lombardi, M.J., & Pagano, P. (2010). *The Impact of Monetary Policy Shocks on Commodity Prices*. Brussels: European Central Bank, Working Paper No. 1232.

Brunnermeier, M.K., & Sannikov, Y. (2013). *A Macroeconomic Model with a Financial Sector*. Princeton, NJ: Princeton University, Department of Economics.

Carroll, R., & Prante, G. (2012). *Long-Run Macroeconomic Impact of Increasing Tax Rates on High-Income Taxpayers in 2013*. London: Ernst & Young.

CBO (2008). *Technological Change and the Growth of Health Care Spending*. Washington, DC: U.S. Congress, Congressional Budget Office, January.

Deising, P. (1962). *Reason in Society*. Urbana, IL: University of Illinois Press.

Edgeworth, F.Y. (1897). The pure theory of taxation. *The Economic Journal*, 7 (25), 46–70.

Edwards, W. (1954). The theory of decision making. *Psychological Bulletin*, 51 (4), 380–417.

Feldstein, M. (1982). Inflation, capital taxation and monetary policy, in R.E. Hall, Ed., *Inflation: Causes and Effects* (pp. 153–168). Chicago: University of Chicago Press.

Ferede, E., & Dahlby, B. (2012). The impact of tax cuts on economic growth: Evidence from the Canadian provinces. *National Tax Journal*, 65 (3), 563-594.

Gali, J., & Gertler, M. (2007a). *Macroeconomic Modeling for Monetary Policy Evaluation*. Cambridge, MA: National Bureau of Economic Research, Working Paper No. 13542.

Gali, J., & Gertler, M. (2007b). Macroeconomic modeling for monetary policy evaluation. *Journal of Economic Perspectives*, 21 (4), 25–45.

Gladwell, M. (2000). *Tipping Point: How Little Things Can Make a Big Difference*. Boston: Little Brown.

Gladwell, M. (2005). *Blink: The Power of Thinking Without Thinking*. Boston, Little Brown.

Gladwell, M. (2008). *Outliers: The Story of Success*. Boston: Little Brown.

Hammond, K.R., McClelland, G.H., & Mumpower, J. (1980). *Human Judgment and Decision Making*. New York: Hemisphere/Praeger.

James, S. (2009). *Incentives and Investments: Evidence and Policy Implications*. Washington, DC: The World Bank.

Jones, E. E., & Harris, V. A. (1967). The attribution of attitudes. *Journal of Experimental Social Psychology*, 3 (1), 1–24.

Kahneman, D. (2011). *Thinking Slow and Fast*. New York: Farrar, Straus and Giroux.

Kahneman, D., & Tversky, A. (1979). Prospect theory: An analysis of decision under risk. *Econometrica*, 47 (2): 263–291.

Kahneman, D., Slovic, P., & Tversky, A. (1982). *Judgment Under Uncertainty: Heuristics and Biases*. Cambridge, UK: Cambridge University Press.

Kaplan, S., & Garrick, B.J. (1981). On the quantitative definition of risk. *Risk Analysis*, 1 (1), 11–28.

Keeney, R.L., & Raiffa, H. (1976). *Decisions With Multiple Objectives: Preference and Value Tradeoffs*. New York: Wiley.

Kelley, H.H., & Michela, J.L. (1980). Attribution theory and research. *Annual Review of Psychology*, 31, 457–501.

Klein, G. (2003). *Intuition at Work: Why Developing Your Gut Instincts Will Make You Better at What You Do*. New York: Doubleday.

Mankiw, N.G., Weinzierl, M., & Yagan, D. (2009). Optimal taxation in theory and practice. *Journal of Economic Perspectives*, 23 (4), 147–74.

Mintzberg, H. (1975). The manager's job: Folklore and fact. *Harvard Business Review*, July-August.

Mirrlees, J.A. (1971). An exploration of the theory of optimal income taxation. *The Review of Economic Studies*, 38 (2), 175-208.

Park, H., Clear, T., Rouse, W.B., Basole, R.C., Braunstein, M.L., Brigham, K.L., & Cunnigham, L. (2012). Multilevel simulations of healthcare delivery systems: A prospective tool for policy, strategy, planning and management. *Journal of Service Science*, 4 (3), 253–268.

Peterson, P.G. (2005). *Running on Empty*. New York: Picador.

Piketty, T. (2014). *Capital in the Twenty-First Century*. Cambridge, MA: Belknap Press.

Piketty, T., & Saez, E. (2012). *A Theory of Optimal Capital Taxation*. Cambridge, MA: National Bureau of Economic Research, Working Paper No. 17989.

Ramsey, F.P. (1927). A contribution to the theory of taxation. *The Economic Journal*, 37 (145), 47–61.

Rouse, W.B. (2007). *People and Organizations: Explorations of Human-Centered Design*. New York: Wiley.

Rouse, W.B. (2009). Engineering Perspectives on Healthcare Delivery: Can We Afford Technological Innovation in Healthcare? *Journal of Systems Research and Behavioral Science*, 26, 573–582.

Rouse, W.B. (2010a). Impacts of healthcare price controls: Potential unintended consequences of firms' responses to price policies, *IEEE Systems Journal*, 4 (1), 34–38.

Rouse, W.B. (Ed.). (2010b). *The Economics of Human Systems Integration: Valuation of Investments in People's Training and Education, Safety and Health, and Work Productivity*. New York: John Wiley.

Rouse, W.B., & Cortese, D.A. (Eds.). (2010). *Engineering the System of Healthcare Delivery*. Amsterdam: IOS Press.

Rouse, W.B., & Serban, N. (2014). *Understanding and Managing the Complexity of Healthcare*. Cambridge, MA: MIT Press.

Sage, A.P., & Rouse, W.B. (2011). *Economic System Analysis and Assessment*. New York: Wiley.

Simon, H. (1972). Theories of bounded rationality, in C.B. McGuire & R. Radner, Eds., *Decision and Organization* (Chapter 8). New York: North Holland.

Slovic, P., Fischhoff, B., & Lichtenstein, S. (1982). Why study risk perception? *Risk Analysis*, 2 (2), 83–93.

Summers, L.H. (1981). Taxation and corporate investment: A *q*-theory approach. *Brookings Papers on Economic Activity*, 1, 67-140.

Tversky, A., & Kahneman, D. (1974). Judgment under uncertainty: Heuristics and biases. *Science*, 185 (4157) 1124–1131.

Tversky, A., & Kahneman, D. (1992). Advances in prospect theory: Cumulative representation of uncertainty. *Journal of Risk and Uncertainty*, 5, 297–323.

Von Neumann, J., & Morgenstern, O. (1944). *Theory of Games and Economic Behavior*. Princeton, NJ: Princeton University Press.

7

SOCIAL PHENOMENA

INTRODUCTION

Table 7.1 lists the social phenomena of interest for the six archetypal problems introduced in Chapter 2 and discussed in Chapter 3. All of these phenomena are concerned with social systems, values and norms, and information sharing. This chapter addresses the state of knowledge in representing these phenomena.

The primary emphases in the chapter are on organizations as social systems and cities as social systems. We will consider social networks and communications, as well as social values and norms. Two contrasts are of particular importance. One is the notion of emergence versus design. The other is direct versus representative political phenomena.

Emergent versus Designed Organizational Phenomena

Emergent organizational phenomena include group formation, consensus building, and social networks. Behaviors of interest include identifying potential group members, communicating among members, and coalition formation. Relevant models include decision theory and game theory, as

Modeling and Visualization of Complex Systems and Enterprises:
Explorations of Physical, Human, Economic, and Social Phenomena, First Edition. William B. Rouse.
© 2015 John Wiley & Sons, Inc. Published 2015 by John Wiley & Sons, Inc.

TABLE 7.1   Social Phenomena Associated with Archetypal Problems

Problem	Phenomena	Category
Traffic control	Drivers information sharing on routes and shortcuts	Social, information sharing
Urban resilience	Peoples information sharing on perceptions, expectations, and intentions	Social, information sharing
Counterfeit parts	Social system of defense industry	Social, organizations
Healthcare delivery	Social system of healthcare industry	Social, organizations
Urban resilience	Social system of city, communities, and neighborhoods	Social, organizations
Counterfeit parts	Values and norms of defense industry	Social, values and norms
Healthcare delivery	Values and norms of healthcare industry	Social, values and norms
Urban resilience	Values and norms of communities and neighborhoods	Social, values and norms

well as network representations of relationships among people. Agent-based models are often employed to create simulations where emergent behaviors can be demonstrated and evaluated.

Designed organizational phenomena include those specified by organizational roles, relationships, and positions. Behaviors of interest include delegation, reporting, and supervision. Hierarchical models are common as are work breakdown structures that assign different elements of work to varying parts of the hierarchy. Heterarchical models can be used to represent workflows at any level of the hierarchy. Workflow models represent how work is accomplished, while the work breakdown structure specifies what is to be accomplished.

Direct versus Representative Political Phenomena

Direct political phenomena include affiliation and choice in small groups or teams. Engagement, preferences, coalitions, and negotiations are relevant specific phenomena. Game theory and social choice theory provide formal models for these phenomena. Social network modeling and agent-based models are also applicable to situations where actors' behaviors directly affect outcomes.

Representative political phenomena are associated with constituencies, interests, coalitions, and negotiations. Communications, polling, and gerrymandering of support can be central when direct choice is not possible. Game theory and social choice theory can be employed here as well, but at a different level of abstraction. Statistical models of interests and preferences, including political affiliations, are often employed.

Behavioral and social aspects of political phenomena include individual and groups responses to socioeconomic situations, possibly involving conflicts and crises. This includes emergence of "outcaste" groups, as well as subsequent formation of new "castes." The dynamics of group formation and emergence of associated values and norms, leading to sanctioned belief systems, play a central role in several important large-scale social phenomena.

Modeling Complex Social Systems

As discussed in Chapter 3, Harvey and Reed (1997) provide a compelling analysis of levels of abstraction of social systems versus viable methodological approaches. Table 7.2 summarizes the results of their analyses, translated into the vocabulary of this book. The appropriate matches of levels of abstraction and viable approaches are highlighted in black.

At one extreme, the evolution of social systems over long periods of time is best addressed using historical narratives. It does not make sense to aspire to explain history, at any meaningful depth, with a set of equations. At the other extreme, it makes great sense to explain physical regularities of the universe using predictive and statistical modeling.

Considering the hierarchy of phenomena introduced in Chapter 1 (Figure 1.1), the types of problems addressed in this book often require addressing physical, human, economic, and social phenomena. Consequently, we need to operate at more than one level of Table 7.2. This implies that our overall representation of the problem – either visually or computationally – will involve multiple types of representations. This poses many challenges that are addressed in Chapters 8–10.

Example – Earth as a System

Looking at the overall system that needs to be influenced can facilitate addressing the challenges of climate change and likely consequences. As shown in Figure 7.1, the Earth can be considered as a collection of

TABLE 7.2   Hierarchy of Complexity versus Approaches (Adapted from Harvey & Reed, 1997)

Hierarchy of Complexity in Social Systems	Predictive Modeling	Statistical Modeling	Iconological Modeling	Structural Modeling	Ideal Type Modeling	Historical Narratives
Evolution of social systems over decades, centuries, and so on.						■
Competition and conflict among social systems						■
Cultural dominance and subcultural bases of resistance					■	■
Evolution of dominant and contrary points of view					■	■
Interorganizational allocations of power and resources				■	■	■
Personal conformity and commitment to roles and norms			■	■	■	■
Intraorganizational allocation of roles and resources		■	■	■	■	■
Distribution of material rewards and esteem		■	■	■	■	
Division of labor in productive activities		■	■	■	■	

TABLE 7.2 *(Continued)*

Hierarchy of Complexity in Social Systems	Predictive Modeling	Statistical Modeling	Iconological Modeling	Structural Modeling	Ideal Type Modeling	Historical Narratives
Social infrastructure of organizations	■	■	■	■		
Ecological emergence of designed organizations	■	■	■	■		
Ecological organization of living phenotypes	■	■	■			
Evolution of living phenotypes	■	■	■			
Regularities of physical universe	■	■				

different phenomena operating on different time scales (Rouse, 2014a). Loosely speaking, there are four interconnected systems: environment, population, industry, and government. In this notional model, population consumes resources from the environment and creates by-products. Industry also consumes resources and creates by-products, but it also produces employment. The government collects taxes and produces rules. The use of the environment is influenced by those rules.

Each system component has a different associated time constant. In the case of the environment, the time constant is decades to centuries. The population's time constant can be as short as a few days. Government's time constant may be a bit longer, thinking in terms of years. Industry is longer still, on the order of decades. These systems can be represented at different levels of abstraction and/or aggregation. A hierarchical representation does not capture the fact that this is a highly distributed system, all interconnected. It is difficult to solve one part of the problem, as it affects other pieces. By-products are related to population size, so one way to reduce by-products is to moderate population growth. Technology may help to ameliorate some of the by-products and their effects, but it is also possible that technology could exacerbate the effects. Clean technologies lower by-product rates but tend to increase overall use, for instance.

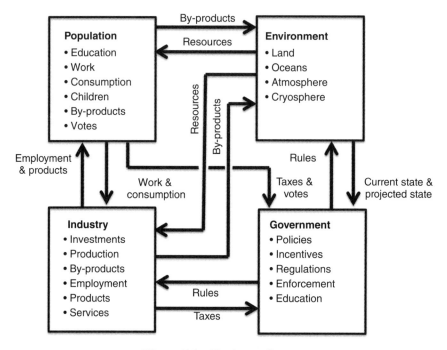

Figure 7.1   Earth as a System

Sentient stakeholders include population, industry, and government. Gaining these stakeholders' support for such decisions will depend upon the credibility of the predictions of behavior, at all levels in the system. Central to this support are "space value" and "time value" discount rates. The consequences that are closest in space and time to stakeholders matter the most and have lower discount rates; attributes more distributed in time and space are more highly discounted. These discount rates will differ across stakeholders.

People will also try to "game" any strategy to improve the system, seeking to gain a share of the resources being invested in executing the strategy. The way to deal with that is to make the system sufficiently transparent to understand the game being played. Sometimes, gaming the system will actually be an innovation; other times, prohibitions of the specific gaming tactics will be needed.

The following three strategies are likely to enable addressing the challenges of climate change and its consequences.

- *Share information.* Broadly share credible information so all stakeholders understand the situation.

- *Create incentives*. Develop long-term incentives to enable long-term environmental benefits while assuring short-terms gains for stakeholders.
- *Create an experiential approach*. Develop an interactive visualization of these models to enable people to see the results.

An experiential approach can be embodied in a "policy flight simulator" that includes large interactive visualizations that enable stakeholders to take the controls, explore options, and see the sensitivity of results to various decisions (Rouse, 2014b). This approach is elaborated in Chapter 8.

PHYSICS-BASED FORMULATIONS

Physicists have addressed a range of social phenomena using models initially developed in statistical physics (Castellano et al., 2009). In these endeavors, humans in social systems are seen as analogous to particles in physical systems. Castellano and his colleagues outline the overall motivations of this research in terms of answering the question, "How do the interactions between social agents create order out of an initial disordered situation?"

They readily agree that the typical assumptions made about humans as microscopic particles ignore the complexity of real individual humans. However, the macroscopic properties of a very large set of humans appear to be largely independent of assumed properties of the microscopic actors. The validity of this conclusion is central to being able to avoid being concerned about realistic models of human behavior.

Building on ferromagnetic phenomena, they consider how charges among nearest neighbor particles tend to align. This tends to lead to ordered states, but thermal noise can disrupt the emergent order. Above a critical temperature, no order emerges. They use this phenomenon as a guide to finding analogous critical variables in social systems.

The topology of social networks is expressed in terms of "degree distributions" or the frequency of the number of connections per node. Research in this area focuses on how the characteristics of the degree distributions affect the order that emerges. For example, they explore the conditions under which the resulting order is unimodal versus multimodal.

They review social dynamics models. Such models usually start with probabilistic transition rates between configurations, leading to a "master equation" for the dynamics of the system.

$$dP(m, t)/dt = \sum [T(m', m)P(m', t) - T(m, m')P(m, t)] \qquad (7.1)$$

where T is a transition matrix whose cells are the probabilities of transitions from one state to another, and the sum is over all m'. This is a gain–loss equation for $P(m)$, which is usually solved numerically. The overall formulation suffers from a lack of empirical data that can be used to estimate parameters.

They also summarize research on agent-based models that emerged from work by von Neumann and others in the 1940s on cellular automata. Agents tend to be rather richer than particle models, but still usually involve dramatic simplifications. They give particular attention to Axelrod's (1997) model of the persistence of diversity. This model assumes that interactions are governed by the degree of similarity of traits between agents. For a small number of traits, complete agreement is reached. For a large number, islands of agreement emerge that represent polarization among groups.

This extensive review considers the following clusters of phenomena:

- Dynamics of opinions – voter models, majority rule models, social impact theory, small group effects, and bounded confidence.
- Cultural dynamics – group formation, consensus, and polarization.
- Language dynamics – "language games" that assume successful communicators are more likely to reproduce than poor communicators.
- Crowd dynamics – flocking, pedestrian behavior, applause dynamics, and spectator waves at sporting events.
- Formation of hierarchies – for densities below some threshold, societies remain egalitarian; above that threshold, hierarchies and large wealth disparities emerge.
- Human dynamics – response times to emails, social networks.
- Social spreading phenomena – diffusion of rumors and news.
- Coevolution of states and topology – evolving networks rewiring themselves.

This enormous body of work, mainly by physicists, tends to be abstract, context free, and difficult to validate. Galam's (2012) treatise on "sociophysics" provides a personal perspective on these issues. He reviews his 30 years of research into "humans behaving like atoms." He considers how and when and statistical physics can be used to understand social phenomena, and discusses a wide range of models.

Galam outlines a set of guidelines for sociophysics. He argues that the field should remain an endeavor of physicists only. This is necessary, he asserts, for the field to grow as a solid science. Not surprisingly, then, he ignores work done by the social sciences and economics. His emphasis is purely on using physics-based mathematical formulations to replicate behaviors that social systems are known to exhibit. The idea that the wealth of underlying assumptions may not be valid does not appear to be a concern. One cannot help but sense a bit of hubris in the whole endeavor.

Example – Castes and Outcastes

We seem to be awash in conflicts today: patriots and terrorists, haves and have-nots, and red states and blue states. One cannot help but wonder where it will all lead. Will all these conflicts work themselves out, or will increasing portions of the economy be devoted to international security and counterterrorism, as well as remediation of the perceived sources of conflict? Alternatively, can destructive conflict be transformed into healthy competition? Where does cooperation fit in these pictures?

Reflecting on these questions brought to mind a classic book by Richard Niebuhr, *The Social Sources of Denominationalism* (1929). Niebuhr's theory of denominationalism can be summarized, albeit somewhat simplistically, in the pithy phrase, "Castes make outcastes, and outcastes make castes." This statement captures, for me, the dynamics of an important source of conflict. As social networks (Burt, 2000) become stronger, there are inevitably people who feel – or are made to feel – that they do not belong and thus they depart. They become outcastes.

This process tends to be quite subtle although, through the centuries, people's abilities to suppress outcastes have certainly had extremes, for example (Foucault, 1995). Nevertheless, it is seldom that people lose their club membership cards and cannot get through the door. Instead, the increased cohesiveness and economic cooperation of the "in group," and their consequent prosperity (Granovetter, 2005), result in people being excluded because they do not believe, cannot compete, or feel uncomfortable with the trappings of the strong dominant social network.

Thus, there are some people who are cast out, and many more people who never have belonged and who consequently become increasingly distant from the mainstream. The members of this collection of outcastes eventually affiliate with each other and seeds of cooperation are sown. Niebuhr suggests that the pursuit of social justice is a driving force of this cooperation – the outcastes feel they have been wronged by the castes. Consequently, the Pilgrims, for example, set sail for North America in the early 17th century.

The affiliation of outcastes leads, over time, to the emergence of a new caste, which via economic cooperation brings prosperity with the strengthening of the new social network. The values of the new caste eventually morph to salvation rather than social justice, as they seek assurance that their increasing wealth is justified and blessed. The success of the new caste inevitably leads to some people feeling excluded and becoming outcastes. Thus, the casting out of those who do not fit in continues, providing fertile ground for the emergence of new castes.

The Pilgrims left England to seek religious freedom. Once settled in Massachusetts, they created an increasingly strong social system and eventually achieved economic success. They came to be intolerant of beliefs that did not fit into their system. Consequently, their Massachusetts caste began to create outcastes. These outcastes eventually migrated to what would become Rhode Island, seeking religious freedom, just as the Pilgrims sought this in leaving England.

Thus, Niebuhr's theory suggests a growing set of "denominations," as is evident for Christians, Jews, Muslims, Buddhists, and so on. He observes that the more mature castes seek sustained economic success; the less mature seek social justice. There are ample opportunities for conflicts among castes. This brings us to another useful theory.

As discussed in Chapter 6, Edward Jones was a leading thinker in the creation of what is known as attribution theory (Jones & Harris, 1967). Attribution theory is concerned with how people attribute cause to observed behaviors and events. Perhaps the best-known element of this theory is the Fundamental Attribution Error. Put simply, people attribute others' success to luck and their own success to hard work. Similarly, they attribute others' failure to a lack of hard work and their own failure to bad luck.

This phenomenon of attribution sets the stage for castes to misperceive each other. In particular, members of a mature caste may perceive members of an immature caste as indolent, not working hard enough to succeed. At the same time, members of an immature caste may perceive members of a mature caste as blessed by birthright rather than having worked for all they have achieved. These differing perceptions can provide a strong motivation for conflict.

The remainder of this example formalizes the notions introduced in these introductory comments (Rouse, 2007). This formalization enables reaching fairly crisp conclusions that explain a great deal regarding the implications of these two theories.

Castes and Outcastes The number of members in the ith caste, NC_i, is given by

$$NC_i(t + 1) = (1 + BC_i - DC_i)NC_i(t) + NA_i(t) - ND_i(t) \qquad (7.2)$$

where BC_i and DC_i are the birth and death rates, respectively, of the caste, and NA_i and ND_i are the number of arrivals to and departures from the caste due,

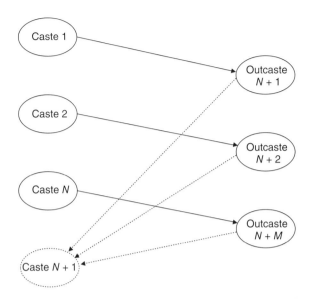

Figure 7.2   Castes Make Outcastes and Outcastes Make Castes

in the former case to recruiting and, in the latter case, to voluntary or forced disaffection, as well as a variety of other causes such as imprisonment, mental health, and so on.

Similarly, the number of members in the jth set of outcastes, NO_j, is given by

$$NO_j(t + 1) = (1 + BO_j - DO_j)NO_j(t) + NA_i(t) - ND_i(t) \qquad (7.3)$$

where BO_j and DO_j are the birth and death rates, respectively, of this set of outcastes, and NA_j and ND_j are the number of arrivals to and departures from the caste.

The process of castes spawning outcastes is depicted in Figure 7.2. Figure 7.2 indicates that not all outcastes are the same. Outcaste Africans and Asians are unlikely, if only due to geography, to belong to the same outcaste population. Thus, at a particular point in time, there are N castes and M sets of outcastes.

Assuming that departures constitute a fraction of the size of a caste, denoted by PD_i for the ith caste, equation (7.2) becomes

$$NC_i(t + 1) = (1 + BC_i - DC_i - PD_i)NC_i(t) + NA_i(t) \qquad (7.4)$$

Departures from sets of outcastes to form a new caste require a critical mass of similar outcastes, a tendency for outcastes to affiliate with each other and, consequently, the emergence of a new caste. Thus, the emergence of a new caste from the jth set of outcastes relates to the number of outcastes NO_j and the probability of members of this set of outcastes affiliating, PA_j. It could reasonably be argued that the probability of a new caste, denoted by k, being formed, PN_k, might be approximated with

$$PN_k = (1 - e^{-\lambda PA_j NO_j}) \qquad (7.5)$$

where λ provides a means for calibrating equation (7.5) to yield reasonable results.

As shown in Figure 7.2, one or more of the sets of outcastes can form a new caste. The dotted lines indicate that the $N + 1^{st}$ caste can emerge from any set of outcastes. While it is certainly possible that two or more set of outcastes can form a single new caste, this would seem unlikely.

Equations (7.2)–(7.5) and Figure 7.2 define the dynamics whereby castes grow, outcastes depart, and new castes are formed. Castes are constantly shedding members and these outcaste members eventually form and/or affiliate with newly emerging castes.

Conflict and Competition   Members of a given caste cooperate and subsequently prosper. Mature castes compete with one another as we see with developed economies in the Americas, Asia, and Europe. Their ensuing battles are often economic, but not necessarily. In contrast, immature castes often conflict, both with each other and with mature castes. This can be understood using Niebuhr's theory of denominationalism.

The typical focus of immature castes is social justice. They feel wronged by the mature castes from which they were outcaste and seek remediation of their grievances. The cooperation of members of the immature castes in pursuit of this cause eventually leads to economic success. As their economic success grows, they eventually shift focus from social justice to salvation, seeking assurance that their economic success is justified and blessed. Thus, they shift from seeking change to seeking preservation of the status quo.

Jones' attribution theory provides a basis for understanding the conflicts that emerge between mature and immature castes. The fundamental attribution error causes mature castes to view the lack of success of members of immature castes to be due to their lack of effort. Similarly, they view their own success as due to hard work, not just luck in belonging to a mature caste. In contrast, members of the immature caste view their lack of success as due to the oppression by the mature caste, the success of whose members is due to accidents of birth.

Of particular importance, while the mature castes attempt to maintain the status quo, immature castes try to foster change. They do not want things to stay the way they are. They want social justice in terms of redistribution of power and resources. Thus, the clash between mature and immature castes seems inevitable.

One might expect castes to move inexorably to maturity, M_C, perhaps following

$$M_C(t + 1) = \alpha M_C(t) + (1 - \alpha)G_E(t) \tag{7.6}$$

As α increases, the maturity (or immaturity) of a caste has high inertia. As α decreases, caste maturity tracks global economic growth, G_E. An elaboration of equation (7.6) would probably also include local (nonglobal) drivers of economic growth. Unfortunately, monotonically growing maturity is not guaranteed, for example, see Lewis (2003).

Model Behavior Equations (7.2)–(7.6) might be computationally modeled as a bifurcating dynamic system (Hideg, 2004) that spawns new systems, or as an agent-based system (Terna, 1998) that spawns new agents, each of which behaves according to the dynamic processes described earlier. With either approach, we would have to be able to represent the expanding state space as new castes are formed, as well as a potentially contracting state space as sets of outcastes disappear.

However, rather than focus on computationally modeling the set of equations laid out here, we can easily imagine the course of NC and NO under reasonable assumptions. If $BC - DC - PD < BO - DO$, which is typically the case, NO will grow more rapidly than NC. Thus, the outcaste population will, over time, likely surpass the caste population. It will take longer if PD is small, but this will nevertheless happen eventually.

The growth of NO will lead to formation of new castes, sooner for large PA and later for small PA. These castes will certainly be immature. Consequently, conflicts will frequently emerge. If α is low, these castes will quickly reach maturity and conflict will be replaced by competition. If the reverse holds (i.e., high α), then conflict will be long lasting.

This suggests that strategies to minimize conflicts should minimize BO, maximize DO, and/or decrease α. Interestingly, there is strong evidence that economic growth decreases BO. Thus, decreasing α can provide two means of decreasing conflict. In contrast, maximizing DO, naturally or otherwise, often seems to be the consequence of not paying attention to the other two strategies.

Model Extensions A deeper level explanation of conflicts between castes and outcastes can be formulated in terms of needs, beliefs, perceptions, and decisions. Perceived attributes of alternative courses of action influence people's decision making regarding whether or not they support a course of action. There is considerable debate about how decision making happens. Do people analytically weigh the attributes of alternatives? Or, do they react holistically to the pattern of attributes of each alternative? As elaborated in Chapter 6, the answers to these questions are highly dependent on the experience of decision makers and whether these are starting wars, investing in factories, choosing spouses, buying cars, or picking candy bars.

In this example, the focus is on how perceptions are formed rather than how decisions are made. The motivation for this emphasis is quite simple. Focusing on people's perceptions of a course of action and assuring that these perceptions are positive, often by modifying the proposed course of action, usually results in their supporting the course of action designed in this way.

The conventional model represents perceptions as being influenced by a person's knowledge and the information presented to them. Succinctly, people's knowledge gained via education and experience combines with the information available to them, the facts at hand, to yield perceptions.

Viewed very broadly, this model seems quite reasonable. However, the simplicity of this model can lead to inappropriate conclusions. For instance, in the context of this simple model, if people's perceptions are other than desired, two courses of action are possible. One can modify the information available and/or educate them to modify and extend their knowledge. In doing so, your goal would be to "correct" misperceptions.

This perspective can lead one to feel that people do not support the proposed course because they do not understand. For example, if people really understood nuclear engineering, they would know that nuclear power is safe and they would want nuclear power plants. If people really understood climate change, they would know that the action being proposed makes sense.

Obviously, this perspective leads to many unsupported initiatives. To enable change decisions, a more elaborate model is needed. What influences perceptions beyond knowledge or "facts in the head," and information or "facts at hand"? A general answer is that people have a tendency to perceive what they want to perceive. In other words, their a priori perceptions strongly influence their a posteriori perceptions. This tendency is supported by inclinations to gain knowledge and seek information that support expectations and confirm hypotheses as discussed in Chapter 6.

These phenomena and a variety of related phenomena can be characterized in terms of the effects of people's needs and beliefs on their perceptions (Rouse, 1993). This characterization is depicted in the

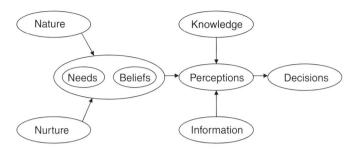

Figure 7.3 Needs–Beliefs–Perceptions Model

Needs–Beliefs–Perceptions Model shown in Figure 7.3. The remainder of this section elaborates this model.

In the context of this model, perceptions that one finds disagreeable are not necessarily incorrect. They are just different. For instance, socialists and capitalists may perceive solutions differently because of differing beliefs about the nature of people and the role of government. As another example, scientists and people who are not technically oriented may have differing perceptions about the impact of technology because of different needs and beliefs concerning the predictability of natural phenomena, desirability of technical solutions, and likely implementations of technologies.

Before considering the state of general knowledge of needs, beliefs, perceptions, and relationships among these constructs, it is useful to note the likely determinants of needs and beliefs. Nature in Figure 7.3 refers to genetic influences on needs and perhaps beliefs. Physical needs, for example, are obviously affected by inherited characteristics and traits. Characteristics such as stature, tendencies for baldness and poor eyesight, and vulnerabilities to particular diseases appear to be hereditary. Inheritance of psychological and social traits is a topic of continuing debate

Nurture in Figure 7.3 refers to the effects of childhood, education, cultural influences, work experience, economic situation, and so on. Clearly, factors such as work experience and economic situation can have a substantial impact on needs. Similarly, education and cultural influences can greatly affect beliefs.

Nature and nurture are included in this model to provide means for affecting needs and beliefs. While nature usually is only affected at the snail's pace of evolution, nurture can be more quickly influenced by modifying situations and hence experiences which, thereby, change needs and eventually beliefs, perceptions, and decisions.

Rouse (1993) shows how this model has been used to address a variety of conflict situations ranging from quality improvement in an automobile company, to disagreements about environmental issues in a semiconductor company, to attempts to transform an electronics company facing loss of their primary market. Surfacing the needs and beliefs underlying conflicting perceptions was found to be the key to making progress on these issues.

Implications Beyond offering a possible explanation for why the world is inherently rife with conflicts, the line of reasoning just outlined can provide important insights into how we should think about diverse social systems. This formulation provides an archetype of a system of systems where the component systems have inherently conflicting objectives. The overall system includes all of humanity and the component systems include the castes and outcastes.

In a corporate setting, castes might be functional areas such as marketing and engineering. Another illustration is an established technology corporation as a caste, and small technology start-ups as the result of outcastes departing. The key is to understand the underlying basis for conflicts between mature and immature castes, appreciate that such conflicts are natural, and devise strategies to move castes beyond these conflicts.

Thus, systems of systems that involve social systems often have inherently conflicting objectives across component systems. These conflicts cannot be designed out of the overall system. Instead, incentives and inhibitions need to be designed to foster creative competition rather than destructive conflict. The model elaborated here would suggest that providing all castes the benefits of economic growth might be an excellent core strategy.

NETWORK THEORY

Network theory is a close relative of physics. Network theory is an area of electrical engineering and computer science and that draws upon graph theory. Network theory is concerned with the study of graphs as a representation of either symmetric or asymmetric relations between discrete entities. Graphs are composed of nodes and arcs or links, where nodes represent entities and arcs or links represent relationships. In physical systems, it is common for arcs to represent flows while nodes represent transformations.

Batty's recent book, *The New Science of Cities* (2013), epitomizes the application of network theory to social systems. His overall line of reasoning is "Our argument for a new science is based on the notion that to understand place, we must understand flows, and to understand flows we must understand networks. In turn, networks suggest relations between people and places, and thus the central principles of our new science depend on defining relations between the objects that comprise our system of interest."

He continues, "Our second principle reflects the properties of flows and networks. We will see that there is an intrinsic order to the number, size, and shape of the various attributes of networks and thus, in turn, of spaces and places that depend on them, and this enables us to classify and order them in ways that help us interpret their meaning."

These premises lead to Batty's seven laws of scaling:

1. As cities grow, the number of potential connections increases as the square of the population.
2. As cities get bigger, their average real income (and wealth) increases more than proportionately, with positive nonlinearity, with their population.
3. As cities get larger, there are less of them.
4. As a city grows in size around its original seed of settlement, which is usually the marketplace or its point of government, the price of land – rent – and the density of occupation decline nonlinearly with distance or travel cost from its central point or central business district.
5. As cities get larger, interactions among them scale with some product of their size, but interactions decline with increasing distance or travel cost between them.
6. As cities get bigger, their central cores decline in population density and their density profiles flatten.
7. As cities get bigger, they get "greener" in the sense of becoming more sustainable.

He reviews a range of models that predict size of cities, income of population, heights of building, locations of retail outlets, and so on, over time.

Closely related is the work of Bettencourt (2005, 2013). His more recent work (2013) addresses the origins of scaling in cities. He defines urban efficiency as the balance between socioeconomic outputs and infrastructural costs. Supporting empirical data include:

- Logarithm of total road miles versus logarithm population: $R^2 = 0.65$;
- Logarithm gross domestic product (GDP) versus logarithm population: $R^2 = 0.96$.

Thus, there is no lack of empirical support for his theorizing.

Bettencourt makes four overarching assumptions that underlie his analyses:

- *Mixing population.* The city develops so that citizens can explore it fully given the resources at their disposal.
- *Incremental network growth.* Infrastructure networks develop gradually to connect people as they join, leading to decentralized networks.
- *Human effort is bounded.* Average social output per capita is independent of city size.
- *Socioeconomic outputs are proportional to local social interactions.* Cities are concentrations of not just people, but rather of social interactions.

He discusses hierarchical network models of urban infrastructure. "The cost of maintaining the city connected (is) the energy necessary for moving people, goods, and information across its infrastructure networks." He notes that "Energy dissipation scales with population like social interactions."

Defining net urban output to equal social interaction outcomes minus infrastructure energy dissipation, he solves for the maximal net urban output as a function of G, the product of GDP and road volume, both per capita. "Cities may be suboptimal either because they do not realize their full social potential or because they do so in a manner that renders transportation costs too high. In either case, urban planning must take into account the delicate net balance between density, mobility, and social connectivity."

Bettencourt (2005) has also explored spontaneous herding and social structure. He argues that "We are part of heterogeneous networks or graphs, sets of links that connect each one of us to all our acquaintances." He continues, "Information cascades lead to the spontaneous formation of large consensus where there are a priori no individual preferences."

He simulates large context-free networks of nodes that embrace trends by seeing the trends embraced by other nodes, including well-connected hub nodes. This continues in the tradition of exploring the implications of pervasive context-free assumptions. It seems that these researchers want to solve every problem in general, but no problems in particular.

Fowler (2006) presents a large-scale network analysis of cosponsorship networks of legislation in the US House and Senate. He introduces a connectedness metric to make inferences about social distance between legislators.

He employs network analysis to find the shortest distance D_{ij} between each pair of legislators in the cosponsorship network. Connectedness is the inverse of the average of these distances from all other legislators to legislator j

$$C_j = (N - 1)/(D_{1j} + D_{2j} + \cdots + D_{Nj}) \tag{7.7}$$

"Connectedness predicts which members will pass more amendments on the floor, a measure that is commonly used as a proxy for legislative influence. It also predicts roll call vote choice even after controlling for ideology and partisanship."

Also in a political context, Koger (2009) argues that "the building blocks of the lobbying game are relationships, with lobbyists and legislators benefitting from bonds based on familiarity and mutual interests." Using data on contributions from lobbyists to legislators in the 2006 electoral cycle, he identifies key dimensions of this network.

The dependent variables were represented by an $N \times N$ matrix where element i, j of the matrix was the number of lobbyists in common for legislators i and j. There were two matrices, one for the House and one for the Senate with, of course, different N. The independent variables were:

- Voting coincidence
- Committee coincidence
- Same party
- Same state
- Electorally vulnerable
- Number of common donors (nonlobbyist, I assume).

He reports that "legislators are more likely to receive donations from the same lobbyists if they are from the same party (in the Senate), state, or committee; if they are both vulnerable in the next election; and the number of common donors increases the more agreement there is in the voting record of a pair of legislators."

These examples of the use of network theory to understand social phenomena include context in the sense that the graphs portray characteristics of cities or political entities. They use data from these domains to construct these graphs. They then infer properties of these graphs upon which they base predictions of urban growth or voting patterns. This overall approach is much more grounded in reality than some of the physics-based formulations.

GAME THEORY

Game theory concerns decision making by two or more individuals where their choices affect each other. The situation can involve conflict or cooperation. Initially game theorists addressed games where each person's gain equaled the losses of the other people involved. This is termed a zero-sum game. Subsequently, game theory has been extended to apply to a wide variety of decision-making situations.

A classic game is Prisoners' Dilemma. Merrill Flood and Melvin Dresher at RAND formulated this game in 1950; Albert Tucker formalized it in the context of prisoners. Two people have been arrested and are in jail unable to communicate with each other. Lacking enough evidence to convict the pair of the principal charge, the jailers try to get each one to incriminate the other. The payoff matrix is shown in Table 7.3.

The decision each person makes depends on what they expect the other person will decide. If the other person remains silent, then one does better by betraying them. If they choose to betray, then one also does better by betraying them. So, the rational choice is betrayal, even though both remaining silent yields better outcomes for both of them. Versions of this game have been used to study a wide range of social and political situations.

Cournot (1838) is often recognized as first publishing an analysis of a formal game, a special case of a duopoly – a market limited to two producers. In a classic and highly cited paper, Hotelling (1929) considers two producers of a common commodity - and a common price - who both try to adjust their production to maximize profit, with the equilibrium being the simultaneous solution of each producer's optimality conditions – as addressed in Chapter 6. He notes that if one of the producers reduces the price for the commodity, they will gain market share and increased profits.

He then considers a situation where both prices and shipping costs (in terms of distance) vary along a one-dimensional region between two providers and again finds the equilibrium price points as a function of shipping costs to find the equilibrium point from the two sets of optimality conditions, one for

TABLE 7.3 Payoff Matrix for Prisoners' Dilemma

	No. 1 Remains Silent	No. 1 Betrays No. 2
No. 2 remains silent	Each serves 1 year	No. 1 goes free No. 2 serves 3 years
No. 2 betrays no. 1	No. 2 goes free No. 1 serves 3 years	Each serves 2 years

each provider. He notes that if both providers agree to increase prices, that is, enter a price fixing agreement, they can increase profits. Without such an agreement, deviating from the price equilibrium is not of value to either provider.

His attention then shifts to shipping or transportation costs. A provider can gain profit by actions that increase the relative shipping costs of his competitor. This might happen by entering into an agreement with the shipping companies for preferential pricing. Lobbying for discriminatory tariffs is another tactic.

Next he looks at the location of the two competitors along the one-dimensional region. Here one provider wants to be close to the other providers if the majority of the market is on the side away from the competitor because the competitor will then have greater shipping costs. A law that constrains location in the favor of one provider is another tactic for gaining competitive advantage.

The publishing of the *Theory of Games and Economic Behavior* by Von Neumann and Morgenstern (1944), which considered cooperative games of several players, facilitated broader interest in game theory. As discussed in Chapter 6, this book also provided an axiomatic theory of expected utility, which enabled formulation of problems involving decision making under uncertainty. By the 1950s, scholars in many disciplines began to adopt game theoretic formulations.

Downs's (1957) classic "An economic theory of political action in a democracy" begins with discussion of Arrow's (1951) conclusion that "no rational method of maximizing social welfare can possibly be found unless strong restrictions are placed on the preference orderings of the individuals in society." Arrow's Nobel Prize winning proof of his "impossibility theorem" means that there is no mechanism for making social decisions that can be shown to work in all situations.

He then elaborates his central hypothesis: "Political parties in a democracy formulate policy strictly as a means of gaining votes." Government is "an entrepreneur selling policies for votes instead of products for money." He first briefly examines "a world in which there is perfect knowledge and information is costless." Not surprisingly, if political parties know exactly what citizens want and provide it, while citizens vote exactly as expected to maximize their known utility functions, the decision-making problem is quite straightforward. He does, however, ignore the problem of how to fund the satisfaction of citizens' needs and desires and their willingness to be taxed to provide the requisite funds.

Downs then examines "a world in which knowledge is imperfect and information is costly." He concludes that "Lack of information converts democratic government into representative government, because it forces the central planning board of the governing party to rely upon agents scattered throughout the electorate." He also considers "influencers" and notes that "Influencers create favorable opinions through persuasion."

He argues that "Imperfect knowledge makes the governing party suscep- tible to bribery." They accept money for TV time, propaganda, and precinct captains in return for policies favorable to those providing the bribes. Lack of information makes it easier for citizens to vote based on ideologies rather than knowledge of policies. "Each party invents an ideology in order to attract the votes of those citizens who wish to cut costs by voting ideologically."

Downs then adopts Hotelling's one-dimensional model with the contin- uum representing ideology rather than distance. He argues mathematically that political parties will converge in the center if one assumes a normal distribution of support along the ideological scale. They will diverge if the distribution along this scale is bimodal. This can result in parties alternating in power, resulting in persistent policy swings and little progress. For multiparty systems, this is much less of a problem. On the other hand, multiparty sys- tems often lead to coalition governments that have difficulty creating coherent policies.

Downs concludes that the returns from being a well-informed voter and therefore voting for the "right" party are much less than the costs of becoming informed. On the other hand, successful lobbying requires being very well informed. This requires that lobbyists be well paid. As noted earlier, lobbyists provide politicians information and money in return for support for policies that benefit those hiring the lobbyists.

There is a wide range of applications of game theory. Kollman and col- leagues (2003) provide a sampling of this range. de Mesquita (2006) provides an interesting treatise on game theory, political economy, and the evolving study of war and peace. He argues for an approach that employs noncoopera- tive game theory as the foundational analytic structure, followed by statistical and case study methods to evaluate model-based explanations.

He begins by indicating that "Democracies rarely, if ever, fight wars with each other." War is usually a costly and inefficient choice for a country's cit- izenry, but may be a good choice for leaders, for example, citizens may rally behind the wartime leader. He discusses several game-theoretic analyses that refute typical balance of power arguments, that is, balance promotes peace and imbalance promotes war.

Democracies are selective about the wars they fight and over the past two centuries have won 93% of their wars. Autocratic states are less selective, because they do not require popular support, and have won 60%, probably greater than 50% due to first-strike advantage.

Democracies only pursue wars where they feel they have a significant advantage. Without such advantage, they pursue negotiations. Thus, democracies seldom war with other democracies, because the country with the lesser advantage will pursue negotiations.

Example – Acquisition as a Game

Pennock (2008) addresses the phenomenon of defense acquisition programs being notorious for their cost overruns, schedule delays, and capability shortfalls. One frequently cited cause is the overreliance on immature technology. Despite the well-known risks entailed in the employment of immature technologies, the practice persists. Why would programs pursue a policy that seems to be counterproductive? To understand this situation, Pennock developed and analyzed a mathematical model of a series of acquisition programs, revealing that when differing stakeholder interests come into play, these programs suffer from a classic tragedy of the commons.

The concept of the tragedy of the commons originated with the overgrazing of shared pastures by sheep and cattle. While it would be best solution if farmers cooperated and limited the number of animals each grazed so as not to destroy the common pasture, each farmer is incentivized to deviate. If only one farmer grazes a few additional animals, he or she will come out ahead without any major adverse consequences for the pasture. But if one farmer deviates, then all farmers will deviate. The tragedy is that the farmers will continue to overgraze their herds even when they understand the consequences of doing so. Government system development programs are no different. When a system development program is shared by multiple, independent groups of stakeholders, each is incentivized to "overgraze."

To support this assertion, Pennock developed a mathematical model of an acquisition program that depends on the maturation of multiple technologies. The key decision variable is the capability improvement goal(s) selected for the program with the assumption that more aggressive goals will require more immature, hence risky, technologies. This model is examined for two different decision models: classical optimization to represent a hypothetical central planner and game theory to represent a group of independent stakeholders with differing interests. What is found is that when there are multiple stakeholders with differing interests, a classic tragedy of the commons emerges.

There are two interrelated difficulties underlying this phenomenon. First, there is a problem with multiple stakeholders each attempting to get their respective requirements addressed by the system. Second, stakeholders tend to be concerned that the acquisition program will take so long that they need to demand as much capability as possible to make it worth the wait.

If the capability improvements targeted are large, these requirements will necessitate the selection of immature technologies. Consequently, these technologies must be matured as part the acquisition program. Developing immature technologies adds to the duration of the program. If there are multiple immature technologies required, then all of them need to be developed for the acquisition program to meet its capability requirements.

Fortunately, these efforts can often be done in parallel. To simplify the analysis, Pennock assumes that there is one technology development effort per capability. The higher the capability improvement goal that is set, the longer and more risky one would expect that technology development effort to be.

Since he was only concerned with the desired level of capability improvement and how long it takes to develop it, he modeled only two "phases": those activities whose durations are impacted by the capability goals and those that are not. The remaining program activities, which are not affected by the capability goals but cannot be done in parallel with technology development, are lumped together into a single term (e.g., engineering and manufacturing development, production, etc.). Since only the total length of the program is of interest, the order of this work, whether before or after technology development, is irrelevant.

The net result is a model of an acquisition program that defines the total duration of a system development program as:

$$P = \text{Max}(X_1, X_2, \ldots, X_N) + S \tag{7.8}$$

where P is the time to complete the program, X_i is the time to complete the technology development associated with capability i, and S is the time required for other activities in the program not associated with technology development. Since technology development is stochastic, each X_i is a random variable governed by a nonnegative distribution function $F_i(x, g_i)$, where g_i is the targeted percent increase in capability i. Pennock quite reasonably assumes that the distributions of the X_i's are independent. As the analysis hinges on the expected length of the program, $E[P]$, any stochastic behavior of S will not materially affect the results of this analysis since it would always be reduced to $E[S]$. So, for simplicity, it is assumed that S is deterministic.

From equation (7.8), it is immediately apparent that the maximization of several random variables will drive the behavior of this model. From probability theory, it is known that the expected value of the maximum of N independent random variables is greater than the maximum of the expected values.

Thus, there is a stochastic interaction effect that results from grouping multiple risky technologies into a single acquisition program. This occurs because as one adds risky activities to a program, the probability that at least one will cause a delay increases.

Pennock solves this model for the case of one decision maker or central planner who is concerned with a series of development efforts and wants to maximize the long-run effective annual growth rate of capabilities. He demonstrates that for an acquisition program that relies on the development of multiple technologies, the optimal policy is to sacrifice some capability improvement aspirations in order to reduce the risk of delays. The resulting optimal capability improvements are denoted by the vector $\underline{G}_C$, where C denotes a single central decision maker. He also shows that as S increases, the elements of $\underline{G}_C$ increase, which diminishes the long-run effective annual growth rate of capabilities.

If it is optimal to target more modest capability improvement goals, thereby enabling selection of more mature technologies, what happens when there are independent stakeholders that have differing interests in the outcome of the acquisition program? In other words, would independent stakeholders agree with the optimal policy, or would they prefer to do something else?

Game theory is the natural approach when concerned with the best response of one person(s) to the actions of others. Pennock modifies the objective function from his mathematical model of an acquisition program, assuming that there is one stakeholder group interested in each technology in the acquisition program. Each of these stakeholder groups only influences the capability goal, g_i, for its technology of interest. Therefore, its best response is to find the optimal value for g_i given the capability policies selected by each of the other stakeholders.

Pennock's analysis seeks an equilibrium point such that no stakeholder has an incentive to deviate. He finds the vector of capability goals, $\underline{G}_E$, where E denotes equilibrium, such that the best response of each stakeholder to that vector is to keep the same capability goal, g_i. If that value for g_i is different than the optimal value found for the central planner, then the stakeholders will each have an incentive to deviate from the optimal policy.

Pennock shows that an equilibrium solution, $\underline{G}_E$, exists for this game. It is more aggressive, that is, larger g_i than the optimal capability goal. The best response of each stakeholder is to deviate from the optimal policy. Consequently, the resulting long-run capability growth rate will be lower than that for the optimal policy.

In essence, this result is a tragedy of the commons because the acquisition program serves as a shared resource to be utilized by stakeholders to achieve their individual objectives. Since there is essentially no cost to each stakeholder (the program is funded with public money), they are incentivized to demand more aggressive capability targets than is optimal. This means the optimal solution is unstable even assuming stakeholders agree to cooperate. Any stakeholder will be better off to push for a little more capability while everyone else follows the cooperative policy. Of course, if one stakeholder deviates, then it is in the best interest of the others to deviate as well. Very quickly they end up at the equilibrium, $\underline{G}_E$.

It is important to note that this is rational behavior. From the perspective of each stakeholder, what each is doing is the best thing that he or she can do. This situation is quite familiar in government acquisition programs. As the expected length of a program increases, more capability is demanded to compensate for the delay. In other words, the new system had better be worth waiting for. Given the characteristics of the technologies involved, the amount of other program time, and the structure of the program, there is an optimal capability policy. However, when there are multiple stakeholders with different mission objectives, the capability policy pursued will likely be more aggressive than optimal. Thus, the effective annual capability growth rate is worse over the long run than one would otherwise expect. A higher growth rate could be achieved if all stakeholders cooperated, and each sacrificed a small amount of capability to speed the completion of the program. This is unlikely, however, because there is always an incentive to deviate.

A third feature of Pennock's model addresses the impact of adding capabilities on the gap between the optimal and equilibrium capability policies. As the number of capabilities with independent stakeholders, N, increases, the gap increases between the optimal, $\underline{G}_C$, and equilibrium, $\underline{G}_E$, policies. As the number of capabilities, N, increases, the optimal solution is to sacrifice a little more capability improvement to achieve a better growth rate over time. However, the opposite is true for the equilibrium policy. When stakeholders do not cooperate, the capability policy becomes more and more aggressive because the interaction effect is exacerbated, and all participants are increasingly worse off.

SIMULATION

A small subset of models of social phenomena can be solved analytically, in the sense of calculating the outputs of a model as a function of its inputs and parameters. When analytic solutions are not tractable, computational solutions are employed and termed simulations. Many simulations are written "from scratch" using various programming languages, or possibly formulated in spreadsheets for, if fortunate, quick and easy execution.

Of particular interest in this book, and elaborated in Chapter 9, are three simulation paradigms. The first of these is system dynamics. The dynamics of systems are often represented as differential equations. There are a variety of means for solving these equations. Most basically, these equations can be represented in electronic circuits, voltages applied to the circuits, and their responses recorded. This is termed analog computing. The first analog computers emerged in the early 1900s, long before digital computers were invented.

By the 1960s, digital computers were rapidly replacing analog computers for solving differential equations, although analog computers retained a speed advantage for some types of problems. Standard software tools for digital simulation were developed, extended, and refined in the past several decades. Of particular note, high-level tools have been created for representation of phenomena in terms of generalized "stocks" and "flows," as will be explained in Chapter 9. Such high-level tools greatly reduce the requirements for programming typical of earlier tools.

Another paradigm is discrete-event simulation. This approach focuses on flows of entities through networks of queues. These networks are usually sufficiently complex to preclude analytical solutions that are possible for standard, simple queues. There are several standard software tools for discrete-event simulation that enable running large numbers of replications to estimate distributions of queue lengths, waiting times, and many other metrics. These tools are discussed in Chapter 9.

The third paradigm is agent-based simulation. Several of the models discussed earlier in this chapter were, at least in effect, agent based. Each agent is modeled as receiving inputs, executing decision rules, and producing outputs. Agents' inputs come from other agents and their outputs go to other agents. The focus with such models is on the macroscopic phenomena that emerge from large numbers of microscopic decisions. Chapter 9 discusses available tools for agent-based modeling.

de Marchi (1999) considers the role of computational modeling in political economics. He indicates that "Computational work springs from a desire to conduct formal, replicable investigations of political phenomena with clearly defined assumptions and hypotheses, typically modeling political actors as locally informed, boundedly rational agents, and possibly with abilities to learn, communicate, and cooperate. Computational modeling is extraordinarily interdisciplinary, and borrows freely from the fields of artificial intelligence, cognitive psychology, economics, information theory, optimization theory, and political science."

He further argues that "Computational modeling is a bridge between the restrictions inherent in analytic approaches, and experimental work that is often not precise in detailing the dynamics of a given phenomenon. Computational models relax many of the more objectionable assumptions in rational choice models, but this does not mean they lack rigor. Rather, computational models, by their very nature, force a researcher to precisely define what rules govern the behavior of political actors and the environments in which they exist. Thus, computational modeling shares a similar epistemological bent with formal theoretic approaches, and should be seen as complimentary and not antagonistic."

There are several trade-offs in developing simulations models revolving around data requirements, computational efficiency, and validation. Voinea's (2007) comparative evaluation of case-based versus agent-based computational models of political terrorism contrasts using rich actual scenarios as the basis for modeling compared to simulations of large numbers of much simpler agents. The former tends to be context rich, while the latter is often rather context free. To avoid having conclusions totally case dependent, several comparable case studies are often needed. This can require significant resources.

Example – Port and Airport Evacuation

Belella and Rooney (2014) present a customized modeling and simulation tool for port and airport evacuation and recovery. This discrete flow modeling and simulation tool includes:

- A road traffic model, overlaying evacuation and recovery vehicles onto existing traffic conditions.
- A pedestrian evacuation model, which establishes the rate at which pedestrians are evacuated from Newark Liberty International Airport facilities to points where they board vehicles or reach other predefined destinations on the airport property.
- A harbor model, evaluating the rate at which Port Newark/Elisabeth Marine Terminals facilities and the Captain of the Port Zone can assume normal operations after an evacuation.

Variations of policies and other parameters can be evaluated in terms of the following measures of effectiveness:

- Time to evacuate Port Newark/Elisabeth Marine Terminals property
- Number of evacuees remaining on site

- Time to evacuate pedestrians
- Time to recover harbor operations
- Traffic congestion during recovery.

To understand the complexity of the problem addressed by this simulation, it is important to indicate that the Port Authority of New York and New Jersey is a joint venture between the States of New York and New Jersey and authorized by the US Congress, established in 1921 through an interstate compact, that oversees much of the regional transportation infrastructure, including bridges, tunnels, airports, and seaports, within the Port of New York and New Jersey. This region includes 1500 square miles (3900 km^2).

Example – Emergence of Cities

Bretagnolle et al. (2003) indicate that "The conception of self-organized systems where a structure observed at a macro-level is supposed to be produced by the interactions between elements at a microlevel has stimulated urban dynamic modeling since the 1980s." Consequently, the overarching premise associated with their model is that "Hierarchical organization and functional differentiation are emergent properties which characterize the level of observation of systems of towns and cities."

They introduce Simpop, an agent-based model designed for simulating the emergence, structuring and evolution of a system of cities, starting from an initial spatial distribution of settlements in a large region, state or set of states, a repartition of resources which can be assessed randomly or exogenously, and rules defining how cities interact, grow and specialize.

They indicate that "The model is a geographic model, in the sense that spatial interactions reflect the power of cities in terms of range of influence of their activities and support for new developments from their access to more or less extended markets. Three types of spatial interactions are the most frequent types of interurban exchanges, linked to different constraints."

"Proximity constrained interactions are representative of many activities for which the distance between supply and demand is an essential constraint. Such interactions are the rule for all central place functions, whatever their level and range, and even if that spatial range is increasing over time. Under that rule, the probability of exchanges are distributed according to a gravity type model."

"Territorially constrained interactions limit a city's influence within boundaries, regional or national. They correspond to all types of administrative or political activities. The interaction rule is modulated according to the nature of the activity, for instance, a capital can levy taxes in an exhaustive way on all cities belonging to its region or state, whereas in the case of other activities, this rule can attribute only a preference for a territorial market."

"Interactions within specialized networks are free from distance constraints, even if exploring them for developing new markets along this line may have differential costs according to the distance. Long-distance trade, maritime transport, a portion of tourism activities, or manufacturing industry, follow this type of spatial interaction rule."

They use Simpop to predict the evolution of a number of Italian cities. They compare their predictions to historical data on the population growth of these cities. The comparison is quite favorable.

URBAN RESILIENCE

The world's population is becoming increasingly urban and many of the largest cities are located near oceans or major rivers. These cities face major environmental threats from hurricanes, typhoons, and storms, the frequencies and consequences of which are aggravated by climate change. As a consequence, there are many major initiatives focused on making cities resilient to such threats (Washburn, 2013).

Before elaborating on the nature of this problem, it useful to briefly review several researchers' perspectives on how many of the approaches discussed in this chapter apply to urban systems. Pumain (1998) discusses sources of complexity in urban systems, as well as theories for modeling complexity, drawing mainly on physics and network theory. Analytical models reviewed include catastrophe theory, predator–prey equations, and master equation models. Spatial and temporal scales for global models of urban structures are discussed. Micro- and macrolevel models are contrasted by reviewing urban microsimulations, cellular automata, multiagent systems, and fractals.

Zhu (2010) provides a review of computational political science. Topics reviewed include text analysis (e.g., sentiment analysis), network analysis (e.g., social networks), twitter feed analysis, and web page analysis, as well as approaches discussed earlier in this chapter such as game theory and agent-based models. The analysis methods are of particular interest as they focus on gleaning insights from empirical measurements in the real world.

Gheorghe et al. (2014) outline a broad field of study they term infranomics. They define this field as follows. "Infranomics is the body of disciplines supporting the analysis and decision making regarding the Metasystem (e.g., the totality of technical components, stakeholders, mindframe, legal constraints, etc. composing the set of infrastructures). Infranomics is the set of theories, assumptions, models, methods and associated scientific and technical tools required for studying the conception, design, development, implementation, operation, administration, maintenance, service supply, and resilience of the Metasystem."

The contributions within this edited collection include many variations of multiattribute analysis of complex infrastructure systems.

Van der Veen and Otter (2002) critique models of urban systems and focus on the phenomenon of emerging spatial structure. Thus, their interest is in levels of aggregation rather than abstraction. They argue that scale and level are often confused. They review microeconomics (consumer and producer behaviors), mesoeconomics (sectors), and macroeconomics (aggregates, aggregate behaviors, and government policy). Within economics, they note that space is often translated, quite simplistically, into transportation costs. Regional economics focuses on were firms and consumes locate themselves. Geography, in contrast, focuses on where things are located and why, with geographic information systems being a primary tool. However, they assert, geographers do not succeed in explaining emergent location behavior.

Byrne (2001) argues for complexity reasoning rather than just modeling and simulation. He discusses the "almost unnerving reliance upon mathematics as the basis of modeling physical phenomena." "This seems even more unnerving when we deal with the natural and social worlds as they intersect. The point is not to dismiss quantification, but to see it as a part of the process of understanding, not as the ultimate representation of understanding."

Relative to many of the models discussed earlier in this chapter, he concludes that "Social structure is more than merely the product of emergent interactions among agents."

Thus, there are advocates and critics of the validity and usefulness of applying the approaches discussed in this chapter to understanding urban resilience as well as guiding policy design and deployment. This sets the stage for a discussion of the full nature of the problem and what we can realistically expect from the methodology elaborated in this book.

A Framework for Urban Resilience

Assuring urban resilience involves addressing three problems. First, there is the *technical problem* of getting things to work, keeping them working, and understanding impacts of weather threats, infrastructure outages, and terrorist acts. The urban oceanography example discussed in Chapter 4 provides a good illustration of addressing the technical problem of urban resilience.

Second, there is the *behavioral and social problem* of understanding human perceptions, expectations, and inclinations in the context of social networks, communications, and warnings. The phenomena and approaches discussed in this chapter, as well as Chapter 5, are relevant to addressing this problem.

Third is the *contextual problem* of understanding how norms, values, and beliefs affect people, including the sources of these norms, values, and beliefs. The needs–beliefs–perceptions model discussed earlier in this chapter is relevant for this problem, augmented by historical assessments of how the neighborhoods and communities of interest evolved in terms of risk attitudes, for example.

Addressing these three problems requires four levels of analysis:

- *Historical narrative.* Evolution of the urban ecosystem in terms of economic and social development – What happened when and why?
- *Ecosystem characteristics.* Norms, values, beliefs, and social resilience of urban ecosystem – What matters now and to whom?
- *People and behaviors.* Evolving perceptions, expectations, commitments and decisions – What are people thinking and how do they intend to act?
- *Organizations and processes.* Urban infrastructure networks and flows – water, energy, food, traffic, and so on – How do things work, fail, and interact?

Returning to the technical problem, the organization and process levels of analysis involve projecting and monitoring urban network flows and dynamics by season and time of day. It requires understanding impacts of weather scenarios such as hurricanes, nor'easters, and heat waves. Also important are impacts of outage scenarios including power loss, Internet loss, and transportation link loss (e.g., bridge or tunnel). Finally, unfortunately, there is a need to understand the impacts of terrorist scenarios, ranging from localized threats like 9/11 to pervasive threats such as might affect the water or food supply.

Issues at the people and behavior level can be characterized in terms of people's questions. At first, their questions include: What is happening? What is likely to happen? What do others think? Later, their questions become: Will we have power, transportation? Will we have food and water? What do others think? Further on, their questions become: Where should we go? How can we get there? What are others doing?

These questions bring us back to the behavioral and social problem. We need to be able to project and monitor answers to people's questions by scenario as time evolves. This involves understanding impacts of content, modes, and frequencies of communications. Integral to this understanding is how best to portray uncertainties associated with predictions of technical problems.

Finally, all of this understanding must occur in the context of *this city's* norms, values, and beliefs, as well as its historical narrative. Thus, we must keep in mind that cities are composed of communities and neighborhoods that are not homogenous. This leads us to the levels of analysis of ecosystem characteristics and historical narrative. These levels of analysis provide insights into who should communicate and the nature of messages that will be believed.

To become really good at urban resilience, we need to address the following research questions:

- How can cities best be understood as a collection of communities and neighborhoods, all served by common urban infrastructures?
- How do policy (e.g., zoning, codes, taxes), development (e.g., real estate, business formation and relocation), immigration, and so on affect the evolution of communities and neighborhoods within a city?
- When technical problems arise, what message is appropriate and who should deliver it to each significantly different community and neighborhood within the city?
- How can we project and monitor the responses of each community and neighborhood to the information communicated, especially as it is shared across social networks?

Consider how the approaches presented in this book can help address these questions. Certainly, empirical studies enable historical narratives and case studies of past events. Historical and anthropological research efforts are central in this regard. They provide the basis for interpreting the context of the city of interest.

Mathematical and computational models of environmental threats, infrastructure functioning, and organizational performance, as well as human response, provide more formal, albeit abstract, explanations of urban events and evolution. Chapters 4–7 have elaborated a wealth of phenomena and approaches for addressing these ends.

Finally, virtual worlds with decision makers in the loop can enable "what if" explorations of sources of problems and solution opportunities. Termed policy flight simulators (Rouse, 2014b), such environments can support decision makers and other key stakeholders to understand issues, share perspectives, and negotiate policies and solutions. Chapters 8–10 address such possibilities in depth.

Summary

Considering the approaches discussed in this chapter, as well as earlier chapters, what can we conclude about our abilities to address urban resilience. First, we must accept that we cannot approach cities in the same ways we address airplanes, factories, and power plants. Cities are laced with too many complex behavioral and social phenomena.

However, we can systematically explore the ways in which cities *might* respond to opportunities, incentives, and inhibitions. We can then identify the conditions more likely to lead to one response rather than another. Then, we can think about how we might engender the conditions leading to more appealing responses. Thus, our focus should be on how to get a city to design itself in ways that enhance resiliency while also improving the quality of life for everyone.

CONCLUSIONS

This chapter began with consideration of the social phenomena identified for the six archetypal problems discussed throughout this book. We then contrasted emergent and designed phenomena, as well as direct versus representative political systems. The overall problem of modeling complex social systems was then considered, with modeling the earth as a system as an example.

Several approaches to modeling social systems were presented, including physics-based formulations, network theory, game theory, and simulations. Examples included castes and outcastes, acquisition as a game, port and airport evacuation, and the emergence of cities. Attention then shifted to urban resilience, including introduction of a framework for understanding the full nature of resilience.

REFERENCES

Arrow, K. J. (1951). *Social Choice and Individual Values*. New York: Wiley.

Axelrod, R. (1997). The dissemination of culture: A model with local convergence and global polarization. *Journal of Conflict Resolution*, 41 (2), 203–226.

Batty, M. (2013). *The New Science of Cities*. Cambridge, MA: MIT Press.

Belella, P.A., & Rooney, B. (2014). A customized modeling and simulation tool for port and airport evacuation and recovery. *Homeland Security Affairs*, 6.

Bettencourt, L.M.A. (2005). *Tipping the Balances of a Small World: Spontaneous Herding and Social Structure*. Santa Fe, NM: Santa Fe Institute.

Bettencourt, L.M.A. (2013). The origins of scaling in cities. *Science*, 340, 1437–1440.

Bretagnolle A., Daudé, E. & Pumain, C. (2003). From theory to modeling: Urban systems as complex systems. *Proceedings of the 13th European Colloquium on Theoretical and Quantitative Geography*, Lucca, Italy, September 5–9, 2003.

Burt, R.S. (2000). The network structure of social capital, in R.I. Sutton & B.M. Staw, Eds., *Research in Organizational Behavior* (Vol. 22). Greenwich, CT: JAI Press.

Byrne, D. (2001). What is complexity science? Thinking as a realist about measurement and cities and arguing for natural history. *Emergence*, 3 (1), 61–76.

Castellano, C., Fortunato, S., & Loreto, V. (2009). Statistical physics of social dynamics. *Review of Modern Physics*, 81 (2), 591–646.

Cournot, Augustin A. (1838). *Recherches sur les Principes Mathematiquesde la Theorie des Richesses. Paris: Hachette*. [Researches into the Mathematical Principles of the Theory of Wealth]. New York: Macmillan, 1897.

de Marchi, S. (1999). *Computational Political Economy: Course Outline*. Durham, NC: Duke University.

de Mesquita, B.B. (2006). Game theory, political economy and the evolving study of war and peace. *American Political Science Review*, 100 (4), 637–642.

Downs, A. (1957). An economic theory of political action in a democracy. *The Journal of Political Economy*, 65 (2), 135–150.

Foucault, M. (1995). *Discipline & Punish: The Birth of the Prison*. New York: Vintage.

Fowler, J.H. (2006). Connecting the Congress: A study of cosponsorship networks. *Political Analysis*, 14, 456–487.

Galam, S. (2012). *Sociophysics: A Physicist's Modeling of Psycho-Political Phenomena: Understanding Complex Systems*. Berlin: Springer-Verlag.

Gheorghe, A.V., Masera, M., & Katina, P.F. (Eds.). (2014). *Infranomics: Sustainability, Engineering Design and Governance*. London: Springer.

Granovetter, M. (2005). The impact of social structure on economic outcomes. *Journal of Economic Perspectives*, 19 (1), 33–50.

Harvey, D.L., & Reed, M. (1997). Social science as the study of complex systems, in L. D. Kiel & E. Elliot, Eds., *Chaos Theory in the Social Sciences: Foundations and Applications* (Chapter 13). Ann Arbor: The University of Michigan Press.

Hideg, E. (2004). Foresight as a special characteristic of complex social systems. *Interdisciplinary Description of Complex Systems*, 2 (1), 79–87.

Hotelling, H. (1929). Stability in competition. *Economic Journal*, 39, 41–57.

Jones, E. E., & Harris, V. A. (1967). The attribution of attitudes. *Journal of Experimental Social Psychology*, 3, 1–24

Koger, G. (2009). *The Beltway Network: A Network Analysis of Lobbyists' Donations to Members of Congress*. Carbondale, IL: Southern Illinois University. Working Paper No. 32.

Kollman, K., Miller, J.H., & Page, SE. (Eds.). (2003). *Computational Models in Political Economy*. Cambridge, MA: MIT Press.

Lewis, B. (2003). *What Went Wrong? The Clash Between Islam and Modernity in the Middle East*. New York: Harper.

Niebuhr, H. R. (1929). *The Social Sources of Denominationalism*. New York: Henry Holt.

Pennock, M.J. (2008). *The Economics of enterprise transformation: An analysis of the defense acquisition system*. PhD Thesis, Georgia Institute of Technology.

Pumain, D. (1998). Urban research and complexity, in C.S. Bertuglia, G. Bianchi, & A. Mela, Eds., *The City and Its Sciences* (Chapter 10). Heidelberg: Physica Velag.

Rouse, W.B. (1993). *Catalysts for Change: Concepts and Principles for Enabling Innovation*. New York: Wiley.

Rouse, W.B. (2007). Two theories that explain a lot: The dynamics of cooperation, conflict and competition. *Insight*, 10 (2), 44–46.

Rouse, W.B. (2014a). Earth as a system, in M. Mellody, Ed., *Can Earth's and Society's Systems Meet the Needs of 10 Billion People?* (pp. 20–23). Washington, DC: National Academies Press.

Rouse, W.B. (2014b). Human interaction with policy flight simulators. *Journal of Applied Ergonomics*, 45 (1), 72–77.

Terna, P. (1998). Simulation tools for social scientists: Building agent based models with SWARM. *Journal of Artificial Societies and Social Simulation*. 1, (2).

Van der Veen, A., & Otter, H.S. (2002). Scale in space. *Integrated Assessment*, 3 (2-3),160–166.

Voinea, C.F. (2007). *A Comparative Review on Computational Modeling Paradigms: A Study on Case-Based Modeling and Political Terrorism*. Bucharest: Analele Universitatii Bucharesti.

Von Neumann, J., & Morgenstern, O. (1944). *Theory of Games and Economic Behavior*. Princeton, NJ: Princeton University Press.

Washburn, A. (2013). *The Nature of Urban Design: A New York Perspective on Resilience*. Washington, DC: Island Press.

Zhu, L. (2010). *Computational Political Science Literature Survey*. State College, PA: College of Information Sciences and Technology, Pennsylvania State University.

8

VISUALIZATION OF PHENOMENA

INTRODUCTION

Not all problems or questions require deep computational exploration. Visual portrayals of phenomena can lead to almost immediate recognition of where a mechanism might fail, or a process flow will lead to bottlenecks, or an incentive system may prompt unintended consequences. In particular, as discussed in Chapter 2, visualization of phenomena can assist in identifying the portions of the overall problem that may merit deep computational modeling. This is central to Step 4 of the overall methodology advanced in this book.

This chapter proceeds as follows. We first briefly address human vision as a phenomenon, primarily to recognize the topic as important but also to move beyond the science of vision to the design of visualizations. We next review the basics of visualization to provide grounding for the subsequent sections. The next section addresses the purposes of visualization, the object of design being the fulfillment of purposes. A visualization design methodology is then presented and illustrated with an example from helicopter maintenance. Visualization tools are then briefly reviewed. Various case studies from Stevens Institute's *Immersion Lab* are discussed. The notion of policy flight simulators is introduced and elaborated. Finally, results are presented from an extensive

Modeling and Visualization of Complex Systems and Enterprises:
Explorations of Physical, Human, Economic, and Social Phenomena, First Edition. William B. Rouse.
© 2015 John Wiley & Sons, Inc. Published 2015 by John Wiley & Sons, Inc.

study of what users want from visualizations and supporting computational tools.

HUMAN VISION AS A PHENOMENON

Human abilities to perceive and recognize patterns are rather amazing and the reason that visualizations are so powerful. An enormous amount of research has been devoted to understanding these abilities. Biederman's (1987) classic theory of human image understanding is a good example of such research. He focused on perceptual recognition of objects, including occluded objects. He derives his theory from five detectable properties of edges in two-dimensional images: curvature, collinearity, symmetry, parallelism, and cotermination. The experimental results presented to support this theory relied on briefly presented images. These results show how human recognition of images is virtually immediate.

Many decades of research has been devoted to designing computer systems that can "see" like humans. Marvin Minsky and Seymour Papert of MIT were pioneers in this area. Their book *Perceptrons* (1969) is recognized as a classic. (As an aside, I audited their course on this topic while I was a graduate student at MIT in the late 1960s and early 1970s.) David Marr, also a faculty member at MIT, posthumously published another classic on computer vision titled *Vision* (1982) that draws upon psychology, physiology, and neuroscience. Minsky later generalized the wealth of findings in this area in *The Society of Mind* (1988).

In general, this topic is concerned with why and how people are able to perceive and understand visualizations. While this topic is fascinating, we cannot approach visualization at this level in this book. It would be akin to trying to understand the phenomenon of humans understanding phenomena. This "meta problem" was noted briefly in Chapter 1, but in this book such a tangent would take us far afield into philosophy rather than engineering.

BASICS OF VISUALIZATION

Edward Tufte is a recognized guru of graphical excellence. In Tufte's (1983) classic, *The Visual Display of Quantitative Information*, he argues that graphical displays should:

- "Show the data
- Induce the viewer to think about the substance rather than about the methodology, graphic design, the technology of graphic production or something else

- Avoid distorting what the data have to say
- Present many numbers in a small space
- Make large data sets coherent
- Encourage the eye to compare different pieces of data
- Reveal the data at several levels, from a broad overview to the fine structure
- Serve a reasonably clear purpose: description, exploration, tabulation, or decoration
- Be closely integrated with the statistical and verbal descriptions of a data set."

Example – Space Shuttle Challenger

In a subsequent book (Tufte, 1997), he provides a powerful illustration of the consequences of ignoring these principles. Tufte reviews the decision to launch the Space Shuttle Challenger on January 28, 1986, resulting in explosion of the Challenger 73 seconds after the rockets were ignited and death of the seven astronauts aboard. This was due to failure of the rockets' O-rings. NASA managers knew about the possible failure of the O-rings. The Morton Thiokol engineers who designed the rocket opposed launching, supporting this recommendation with a 13-chart presentation. However, the data that were presented in the charts were not fully representative of the phenomena of interest, that is, the impact of temperature on the performance of the O-rings. The charts obscured the phenomena and provided ample room for NASA to argue that the evidence did not support a decision to delay the launch. Tufte concludes, "There was a clear proximate cause (of the accident): an inability to assess the link between cool temperature and O-ring damage on earlier flights. Such a prelaunch analysis would have revealed that this flight was at considerable risk." He also notes that, beyond the inadequately represented data, "The accident serves as a case study of groupthink, technical decision making in the face of political pressure, and bureaucratic failures to communicate."

Tufte's guidance is often applied within the broader context of humans interacting with systems, ranging from everyday things (Norman, 1988), to computers and portable devices (Moggridge, 2007), to complex systems and organizations (Rouse, 2007). Moggridge argues that the designing interactions within such contexts should be considered in terms of a hierarchy of complexity regarding humans and their interactions with the world:

- "Ecology: The interdependence of living things, for sustainable design

- Anthropology: The human condition, for global design
- Sociology: The way people relate to one another, for the design of connected systems
- Psychology: The way the mind works, for the design of human–computer interactions
- Physiology: The way the body works, for the design of physical (hu)man–machine systems
- Anthropometrics: The sizes of people, for the design of physical objects"

This characterization of physical, human, and social phenomena has parallels in many of the discussions throughout this book. The key point here is that visualizations are used by humans in the context of interactions with technological systems and other people, and that these interactions occur in the context of economic and social systems that have historical roots and contemporary manifestations. Context, at multiple levels, really matters.

Data visualization has received an enormous amount of attention in recent years. Nevertheless, there are a variety of unsolved visualization problems (Chen, 2005). He argues for increased focus on usability, noting that the growth of usability studies and empirical evaluations has been relatively slow. It is much easier to conclude that a particular visualization is "cool," than to determine that it is both usable and useful.

Chen also asserts that we need increased understanding of elementary perceptual-cognitive tasks, including the notion of visual inference and the impact of prior knowledge on task performance. This, he asserts, is particularly important for visualizations that are less about structure and more about dynamics and thus require visualizing changes over time.

The interpretation of visualizations often requires domain knowledge. For example, the symbology displayed and relationships among symbols often depend on prior understanding of aircraft or power plants, for example. There has been limited compilation of the types of visualizations relevant to different domains, including how to visualize domain knowledge in itself.

He also argues for practical needs such as education and training, intrinsic quality measures, and understanding the scalability of visualizations. For example, what works for 100 nodes may be less useful for 100,000 nodes. Finally, Chen includes the challenge of understanding the esthetics of visualizations. What looks good and what does not and why? The answers to these questions are unlikely to be universal across domains and cultures.

PURPOSES OF VISUALIZATIONS

Beyond being aesthetically appealing, visualizations are usually intended to serve some useful purpose. Indeed, without purposes, visualizations are all just colors and curves. Not surprisingly, it is important to define the purpose of the visualization before, or at least during, its creation.

Ware (2012) suggests the following purposes:

- "Visualization provides an ability to comprehend huge amounts of data
- Visualization allows the perception of emergent properties that were not anticipated
- Visualization often enables problems with the data to become immediately apparent
- Visualization facilitates understanding of both large-scale and small-scale features of the data
- Visualization facilitates hypothesis formation."

Rasmussen (1983, 1986) addressed the role of visualizations in operating and maintaining complex engineered systems such as nuclear power plants. He argued for the merits of thinking about information displays in terms of abstraction and aggregation. He proposed the following abstraction hierarchy:

- "Functional Purpose – production flow models, system objectives
- Abstract Function – causal structure; mass, energy, and information flow topology, etc.
- Generalized Functions – standard functions and processes, control loops, heat transfer, etc.
- Physical Functions – electrical, mechanical, chemical processes of components and equipment
- Physical Form – physical appearance and anatomy, material and form, locations, etc."

Thus, the purpose of a system is a more abstract concept than how it functions. In turn, the systems' abstract and generalized functions are more abstract than its physical functions and form. A later discussion of work by Frey, Rouse, and Garris illustrates the power of thinking in these terms.

Levels of aggregation are best illustrated in terms of the decomposition of a system, for example, an automobile. At the abstraction level of physical form, the vehicle can be decomposed into power train, suspension, frame, and so on. The power train can be decomposed into engine, transmission, drive shaft, differentials, and wheels. The engine can be decomposed into block and cylinders, pistons, camshaft, valves, and so on. As noted, these levels of aggregation are all represented within the same level of abstraction – physical form.

Rasmussen's hierarchies of abstraction and aggregation play a central role in the visualization design methodology presented later in this chapter. These constructs are central to usefully portraying complex systems for management, design, operations, and maintenance. By the way, Rasmussen formalized these ideas, based on many years of research, while on sabbatical with my research group at Georgia Tech in the early 1980s.

As is elaborated next, one can view a use case of a set of visualizations as a trajectory in an abstraction–aggregation space. Each point in this space will have one or more associated visualizations. Each of these visualizations may portray, individually or in combination, data (e.g., performance history of subsystems), structure (e.g., what connects to what), dynamics (e.g., response over time), or other pertinent representations.

Examples – Co-Citation Networks and Mobile Devices

Chen (2004) addresses visualization of a knowledge domain's co-citation network. A co-citation network is comprised of the linkages, via reference lists, among the articles published in a domain of research. He demonstrates that a search for intellectual turning points can be narrowed down to visually salient nodes (i.e., publications) in the visualized network. He uses this result to argue a more general point, "A network visualization with the fewest number of link crossings is regarded as not only aesthetically pleasing but also more efficient to work with in terms of the performance of relevant perceptual tasks." This is a good example of visualizing the structure of a domain for the purpose of understanding the source of particular phenomena in that domain.

Basole (2009) provides another excellent example. He characterizes the mobile device ecosystem "as a large and complex network of companies interacting with each other, directly and indirectly, to provide a broad array of mobile products and services to end-customers." He argues that "with the convergence of enabling technologies, the complexity of the mobile ecosystem is increasing multifold as new actors are emerging, new relations are formed, and the traditional distribution of power is shifted." He "draws on theories of network science, complex systems, interfirm relationships, and the creative art and science of visualization, to identify key players and map the complex structure and dynamics of nearly 7000 global companies and over 18,000 relationships in the converging mobile ecosystem."

Basole's approach "enables decision makers to (i) visually explore the complexity of interfirm relations in the mobile ecosystem, (ii) discover the relation between current and emerging segments, (iii) determine the impact of convergence on ecosystem structure, (iv) understand a firm's competitive position, and (v) identify interfirm relation patterns that may influence their choice of innovation strategy or business models."

DESIGN METHODOLOGY

Computer technology, including graphics hardware and software, has made it possible to create impressive and pleasing visualizations of a wide range of phenomena. If one's sole purpose is to impress people and collect "wows," then the tools are available to achieve these ends. In this book, however, the goal is to create interactive visualizations that serve particular purposes.

Earlier in this chapter, purposes of visualization were discussed. Now, we need to consider users' purposes, which visualizations are intended to support. Users' purposes seldom include using visualizations; these are simply the means to other ends such as:

- *Problem Solving*, perhaps using the topographic rules (Rouse, 1983) discussed in Chapter 5 to explore structural relationships underlying the phenomena of interest.
- *Pattern Recognition*, perhaps using the symptomatic rules (Rouse, 1983) discussed in Chapter 5, or recognition-primed decision making discussed in Chapter 6 (Klein, 2003), to identify regularities and anomalies
- *Procedure Execution*, as illustrated later in this chapter for helicopter maintenance (Frey et al., 1992, 1993), which involves understanding how to execute procedural steps.
- *Navigation*, involving maps and signs (Rasmussen, 1983) needed to move from one location to another, ranging from geographic locations to finding the subsystems of a power plant, for example.

The methodology outlined next is intended to help one to design visualizations that will support users in their pursuits of these types of purposes.

Step 1: Identify information use cases, including knowledge levels of users. Use cases provide descriptions of alternative ways in which users will employ the visualizations to achieve their purposes – *before* the visualizations have been created. These high-level descriptions can be characterized in terms of six general tasks:

- *Retrieve* data and visualizations relevant to questions of interest
- *Recognize* characteristics of interest across chosen attributes
- *Construct* or select and parameterize representations provided
- *Compute* outputs of constructed representations over time
- *Compare* outputs to objectives or across output variations
- *Refine* constructed representations and return to *Compute*

Examples of the use of such descriptions are provided in the following subsection.

Step 2: Define trajectories in abstraction–aggregation space. Use cases define what information is needed and what actions are taken at every step of the task of interest. This includes the levels of abstraction and aggregation for requisite information elements and controls for each task.

Step 3: Design visualizations and controls for each point in space. Transform the outputs of Step 2 into what specifically users can see and do. There is a wealth of possibilities, which need to be compiled into a manageable set of choices. Note that many of the possibilities will be domain dependent.

Step 4: Integrate across visualizations and controls to dovetail representations. Visualizations and controls should not completely change for each step in a task. Integrated visualizations may be able to support more than one step. Individual controls may affect more than one view.

Step 5: Integrate across use cases to minimize total number of displays and controls. The set of visualizations may serve more than one purpose. For example, novices may use it to learn about a domain while experts use it to address real problems of interest. Experts may, for example, see the same central visualizations as novices, but have access to additional information and controls to enable manipulations of phenomena that novices would not understand.

Use Case Illustrations

The initial result of Step 1 is often a rather high-level description, perhaps just a few phrases. Step 2 may initially involve a single point in the abstraction–aggregation space, and hence Step 3 may initially involve a single visualization. Thus, the initial "spiral" through the prototyping and evaluation cycle will likely not involve Steps 4 and 5 at all. The following examples illustrate how successive spirals add increasing substance to the interactive visualizations.

Figure 8.1 Visualization of Hoboken Being Flooded (*Source*: Reproduced with permission from Blumberg (2013). Copyright © 2013, Stevens Institute of Technology.)

Blumberg (2013) developed a demonstration to help people to recognize and understand the nature of the flooding in Hoboken, NJ, associated with hurricanes and other major storms. His portrayal in Figure 8.1 enables them to recognize what is happening to Hoboken – "Where did the water come in and where did it go?" The water levels in this visualization were computed using the urban oceanography model discussed in Chapter 4. In this way, the predicted intensity of the storm could be varied and flood levels predicted accordingly. The general tasks associated with this demonstration include Retrieve and Recognize.

Yu (2014) developed an interactive game that focuses on all the ingredients of a hospital acquisition decision, including what your competitors are doing or are likely to do (Figure 8.2). She provides a range of detailed information intended to support problem solving – "What should we do?" General tasks supported include Retrieve and Recognize for accessing a wealth of financial and operational performance information on New York City's 44 hospitals. The game also supports Construct, Compute, Compare and Refine for configuring and running the 10-year simulation of acquisition and merger offers, acceptances, and rejections. Typically, 12–15 hospitals disappear (as businesses, not as facilities) over the 10-year period, although not always the same set of hospitals remain.

Park and colleagues developed a multilevel simulation to enable Emory University to assess the scalability of its pilot prevention and wellness program from 700 employees to 60,000 covered lives (Park et al., 2012; Rouse

Figure 8.2   Hospital Acquisition Game (Yu, 2014)

& Serban, 2014). The simulation operated on four levels – people, process, organization, and ecosystem.

- *People*: An agent-based representation of program participants in terms of health records and risks of diabetes and heart disease.
- *Process*: A discrete-event representation of the prevention and wellness process as support by participants' "health partners."
- *Organization*: A microeconomic representation of the organization delivering the program in terms of revenue, costs, and health outcomes.
- *Ecosystem*: A macroeconomic representation of economics of prevention and wellness from the perspective of Emory's Human Resources (HR) function.

The agent and process models were defined by risk stratification parameters and treatment decision parameters. Macroeconomic settings (e.g., inflation, discount rate) and microeconomic settings (e.g., program entry requirements and growth rate) affected economic valuation for both the organization and ecosystem levels.

The general tasks supported are Construct (i.e., select and parameterize), Compute, Compare, and Refine. The simulator was used by groups of medical executives with technical support by Park and his colleagues to computationally identify the conditions under which the program would be scalable with

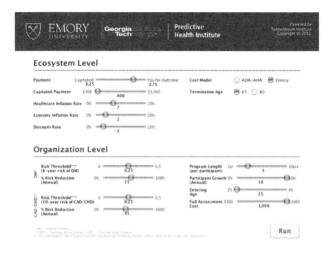

Figure 8.3 Dashboard for Emory Simulator (Park et al., 2012)

positive return on investment (ROI) to Emory. This exploration is elaborated later in this chapter (Figure 8.3).

The *Product Planning Advisor* (Rouse & Howard, 1995; Rouse, 2007) was developed to support new product planning. It supports users to configure a multiattribute utility theory model of market stakeholders' preferences, with independent utility functions for each stakeholder. Users define their planned offerings, as well as competitors' offerings, in terms of functionality and a mapping of functionality to stakeholders' attributes. Expected utility calculations are performed with capabilities to explore sources of differences among offerings. "What if?" capabilities were provided to assess the impacts of improving any selected subset of attributes.

Tasks supported are Construct (i.e., select and parameterize), Compute, Compare, and Refine. Cross-disciplinary groups – engineering, manufacturing, finance, marketing, and so on – used the *Product Planning Advisor*. It was used to explore alternative offerings, markets, and competitive positions. It is important to note that much iteration of prototyping and evaluation was needed before the *Product Planning Advisor* became a successful product, eventually selling a large number of copies. The eventual product interface is shown in Figure 8.4.

The *Technology Investment Advisor* (Rouse et al., 2000; Rouse, 2007) was developed to enable technology companies to "backcast" (rather than forecast) technology investments required to achieve their market objectives. Users defined families of *S*-curves to represent generations of market offerings, for example, versions 1.0, 2.0, and so on. Beyond the *S*-curve for

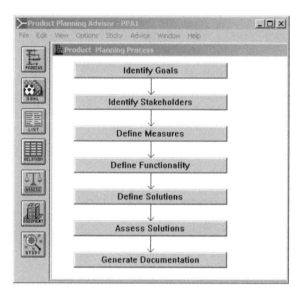

Figure 8.4   Product Planning Advisor

each generation, microeconomic models of product development and market penetration models were defined. The R&D investments necessary to enable these eventual offerings were characterized as real options, where the R&D budget was the option purchase price and the market deployment budget was the option exercise price. Given the very uncertain nature of such endeavors, Monte Carlo analysis across all model parameters was included to assess sensitivity of projections to parameter uncertainties.

Tasks supported by the *Technology Investment Advisor* were Construct (i.e., select and parameterize), Compute, Compare, and Refine. Figure 8.5 includes two of the visualizations associated with this tool. Note that the valuation visualization was the initial conception of this software tool. Prototyping and evaluation quickly led to the conclusion that users wanted a more principled and transparent means to populating the cells of the valuation spreadsheet. This led to the backcasting models and *S*-curves.

EXAMPLE – BIG GRAPHICS AND LITTLE SCREENS

By the early 1990s, two decades before the iPad and other tablet devices, computer-generated displays were getting small and flat enough, and sufficiently lightweight, to entertain using them as portable documentation systems. Under contract to the US Navy, Frey et al. (1992, 1993) researched

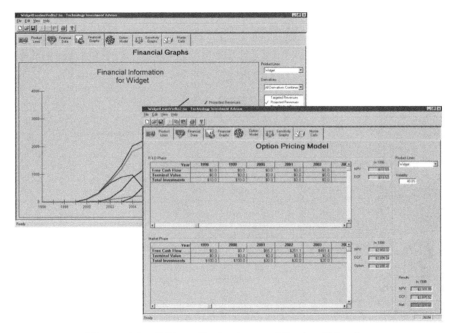

Figure 8.5 *S*-Curve Projections and Option Valuation

this possibility for maintenance of the blade fold system of the SH-3 helicopter.

The standard documentation to support blade fold maintenance included large (11 × 17 inch) location and schematic diagrams. The initial idea was to put these on small screens with capabilities to pan and zoom. We felt, however, this would not take advantage of the display technology. It was easy to imagine users getting lost due to having to zoom in to make the displays readable and getting lost panning over the large, flat geography of the schematics.

There were three tasks that these displays were intended to support. Procedure following involved using fully proceduralized job performance aids that specified each step in the process in terms of what to look at and what to do. A second task was circuit tracing, which mainly involved tracing electrical circuits and locating test points. Most challenging was problem solving. This involved determining the operational symptoms of failures of certain devices, identifying "half-split" test points given failure symptoms, and troubleshooting failures for which there were no procedures or job aids.

As shown in Table 8.1, these tasks were classified as thinking or doing tasks. Thinking tasks involved inference, deduction, interpretation, and decision making. Doing tasks required navigation, locating devices, components or test points, observation, and manipulation.

TABLE 8.1   Effects of Users' Tasks on Use of Abstraction/Aggregation Levels (Based on Frey et al., 1992)

Tasks/Attributes	Change of Representation	Change of Field of View
Thinking • Inferring • Deducing • Interpreting • Deciding	Likely to benefit from more abstraction	Likely to involve movement to less aggregation
Doing • Navigating • Locating • Observing • Manipulating	Likely to require much less abstraction	Likely to involve movement among levels of aggregation

A set of 45–60 computer-generated displays was designed to replace the traditional hard copy schematics. The number of displays made available depended on the experimental conditions, which varied across the five experiments discussed next. These displays can be classified in terms of the three-by-three abstraction–aggregation space shown in Table 8.2. These displays were designed with the hypotheses in Table 8.1 in mind.

Five experiments were conducted involving 55 SH-3 maintenance personnel, specifically Aviation Electricians at the Jacksonville Naval Air Station. These electricians were classified as in terms of experience level, either high or low. The characteristics of the five experiments and their overall results are summarized in Table 8.3.

Key results can be summarized as follows. Thinking tasks benefitted more from the high abstraction displays, that is, high abstraction displays were used much more for problem-solving tasks versus procedure following and circuit tracing tasks. High abstraction displays provided greater benefits to the more experienced maintainers.

High aggregation displays were used most for procedure following tasks. Aggregation levels employed shifted to lower levels as task performance progressed. Lower levels of abstraction were inherently more useful for procedure following and circuits tracing as less abstract reasoning was required.

TABLE 8.2 Abstraction–Aggregation Space (Based on Frey et al., 1993)

Level of Abstraction	Level of Aggregation		
	High	Medium	Low
High (Flows)	Flow diagrams grouped by function	Electrical and hydraulic flow diagrams	None
Medium (Schematics)	Schematic diagrams grouped by function	Function-oriented schematics	Device-oriented schematics
Low (Locations)	Entire helicopter and location of major assemblies	Major assemblies and locations of subassemblies	Subassemblies and locations of components

Some of the maintainers, particularly those with less experience, had some difficulty getting lost in the large numbers of displays in the hierarchy. Inexperienced maintainers also had more difficulty understanding the high abstraction displays. "How to use" training decreased these difficulties, especially for the less experienced personnel.

We have often found that complex interactive visualizations, as well as decision support in general, require training in their use. Not all interactive visualizations are inherently intuitively easy to understand and use. In one application, we had to provide aiding in the use of the aiding. This resulted in support akin to Obi-Wan Kenobi in Star Wars that monitored use of the aiding and suggested different behaviors when users were not taking advantage of the full capabilities of the aiding.

The display set designed for the SH-3 maintainers demonstrably improved their performance. Beyond this impact, the maintainers, their instructors, and their managers asked to keep the experimental displays once the experiments were completed. This required approval, which had its own complications. Nevertheless, we viewed this as a definite endorsement of using the abstraction–aggregation space as a framework for designing interactive visualizations.

VISUALIZATION TOOLS

There is a wide range of software tools for creating visualizations, or at least elements of interactive visualizations. Any exposition on these tools would

TABLE 8.3 Experimental Findings (Based on Frey et al., 1993)

Experiment	Variables	Measures	Subjects	Results
1	Experience Displays	• Number of displays	6	• Fewer displays used with experimental material
2 and 3	Experience Task type	• Time on abstraction and aggregation levels	10 and 13	• Thinking tasks require higher levels of abstraction • Doing tasks require lower levels of abstraction • On thinking tasks, experienced subjects use high abstraction displays more than inexperienced subjects • Aggregation use shifts to lower aggregation as task progresses
4	Experience Training content Displays	• Time to solution • Simulated maintenance performance • Time on abstraction and aggregation levels	16	• "How to use" training improved performance of inexperienced subjects • Inexperienced subjects performed more poorly with high abstraction displays
5	Displays	• Time to solution • Simulated maintenance performance	10	• Performance improved by increasing either abstraction or aggregation over baseline displays

quickly be out of date. Nevertheless, it is useful to briefly review a few of these offerings.

Data

The topic of data visualization has received enormous attention in recent years. The most commonly used visualization tool has been, for many years, Microsoft Excel. Millions of users are familiar with its Charts function. More recently, other popular tools are Google Charts, Tableau, R, and D3, the latter two are open source and, hence, free.

Structure

The visualization of structure has long been of interest, particularly to engineers and other designers. Block diagrams have been used for many decades. These diagrams depict the parts or functions of a system as blocks and the relationships of the blocks as lines between the blocks. Functional flow block diagrams have been used since the 1950s. Block diagram algebras enable simplification of diagrams with many loops.

IDEF, originally meaning ICAM Definition and later Integration Definition, is a family of modeling languages in the field of systems and software engineering. Uses range from functional modeling to object-oriented analysis and design. They are most commonly used in the aerospace and defense industry. This set of tools is in the public domain and, hence, free.

A systemigram (Boardman & Cole, 1996) is a network representation that includes words and phrases on the links between nodes that describe the relationships among the entities in the nodes. Systemigrams are intended to represent the process and activities within an enterprise and the relationships among them. These diagrams are usually much more readable and readily understandable than other forms of structural diagrams.

Decision trees employ a tree-like graph of decisions and their possible consequences, including probabilistic outcomes, resource costs, and utilities of outcomes. For complex decisions situations, decision trees can become unwieldy. This led in the 1970s to the formulation of influence diagrams, a compact graphical and mathematical representation of a decision situation (Howard & Matheson, 1981).

Dynamics

There are several types of diagrams that enable understanding how a system evolves in time. Causal loop diagrams, sometimes seen as variants of influence diagrams, add indications of whether the relationship between two phenomena is positive or negative, denoted by plus or minus signs on the arcs connecting the nodes representing the two phenomena. Closed cycles in the diagram are of particular interest. A loop with all positive signs is called a reinforcing loop, while a loop with at least one negative sign is termed a balancing loop.

Stock and flow diagrams provide richer representations than causal loop diagrams. Stocks represent elements of systems whose values at any given instant in time depend on past behavior of the systems. Flows represent the rate at which stocks are changing at any given instant, they either flow into stocks, thereby increasing them, or flow out of stocks, causing decreases. Other elements of such diagrams are sources and sinks.

Markov diagrams provide a means for representing the propagation of uncertainty in a system. These diagrams represent probabilistic transitions between nodes, where nodes represent the state of the systems, for example, number of entities being processed. A key assumption with these diagrams is that the future state of the system only depends on its current state, not past states. There are variations of Markovian descriptions that avoid this assumption, but analysis of such variations is much more complicated.

These diagrams often provide the basis for developing computational simulations of the interactions, over time, of phenomena in the diagram. Software tools such as Stella and Vensim are popular simulation engines. In some cases, the diagrams are first converted to differential equations (continuous time) or difference equations (Discrete time) and then solved with tools such as Mathematica and MATLAB.

IMMERSION LAB

The wide variety of tools briefly summarized earlier can be used to construct elements of an overall interactive visualization. The issue of how best to integrate these various elements quickly arises when addressing questions of interest in the context of a complex system in a particular domain. Consideration of this issue led to the following observations:

- Many of the phenomena in our critical public–private systems are very complex and becoming more so.
- Many of the key stakeholders in these systems are not technically sophisticated yet they have enormous influence on outcomes.
- These stakeholders can be engaged and influenced by being immersed in the complexity of their domain.

- The Stevens Institute *Immersion Lab* can attract key stakeholders and sponsors to interactive problem-solving and decision-making sessions.
- If the interactive visualizations are well conceived and integrated, many people will report that they did not realize what they experienced was possible.

Figure 8.6 is a photograph of the first working meeting in the *Immersion Lab* in September 2013. The group shown includes the senior staff members from the Antarctica team of the National Science Foundation, personnel from the Lockheed Martin support team, and researchers from Stevens Institute. They are addressing how best to transform McMurdo Station in Antarctica to be more energy and cost efficient while also improving the delivery of the science mission on the continent.

The *Immersion Lab* includes an 8 foot by 20 foot, 180° visualization platform with 7 touch-sensitive displays. These large touch displays can be driven in any combination, ranging from seven independent displays to one large, integrated display. Local display generation computers, linked to nearly supercomputer capabilities, drive the displays.

While Figure 8.6, as well as Figure 8.2, show people demonstrating their multifaceted interactive visualizations to seated audiences, working sessions more often involve small groups (6–8 people) actually immersed in the 180° environment, pursuing various "What if?" explorations. While these small groups often begin seated, the lure of the immersive experience quickly has them absorbed in the interactive environment.

Figure 8.6 Virtual Antarctica

Other examples of *Immersion Lab* projects include the coastal urban resilience project (Figure 8.1) and healthcare ecosystem project (Figure 8.2), as well as several projects concerned with the dynamics of the financial system. Various sponsors and other stakeholders schedule meetings and events in the *Immersion Lab* months in advance.

POLICY FLIGHT SIMULATORS

As we have worked with groups of senior decision makers and thought leaders using interactive visualizations such as those in Figures 8.2 and 8.3, they have often asked, "What do you call this thing?" I suggested "multilevel simulations," but I could tell from the polite responses that this did not really work. At some point, I responded "policy flight simulator" and immediately knew this was the right answer. Numerous people said, "Ok, now I get it." This led to a tagline that was also well received. The purpose of a policy flight simulator is to enable decision makers to "fly the future before they write the check."

Policy flight simulators are designed for the purpose of exploring alternative management policies at levels ranging from individual organizations to national strategy. This section focuses on how such simulators are developed and on the nature of how people interact with these simulators. These interactions almost always involve groups of people rather than individuals, often with different stakeholders in conflict about priorities and courses of action. The ways in which these interactions are framed and conducted are discussed, as well as the nature of typical results.

Background

The human factors and ergonomics of flight simulators have long been studied in terms of the impacts of simulator fidelity, simulator sickness, and so on. Much has been learned about humans' visual and vestibular systems, leading to basic insights into human behavior and performance. This research has also led to simulator design improvements.

More recently, the flight simulator concept has been invoked to capture the essence of how interactive organizational simulations can enable organizational leaders to interactively explore alternative organizational designs computationally rather than physically. Such explorations allow rapid consideration of many alternatives, perhaps as a key step in developing a vision for transforming an enterprise.

Computational modeling of organizations has a rich history in terms of both research and practice (Prietula et al., 1998; Carley, 2002;

Rouse & Boff, 2005). This approach has achieved credibility in organization science (Burton, 2003; Burton & Obel, 2011). It is also commonly used by the military.

Simulation of physics-based systems has long been in common use, but the simulation of behavioral and social phenomena has only matured in the past decade or so. It is of particular value for exploring alternative organizational concepts that do not yet exist and, hence, cannot be explored empirically. The transformation of health delivery is, for example, a prime candidate for exploration via organizational simulation (Basole et al., 2013).

This section focuses on the nature of how people interact with flight simulators that are designed for the purpose of exploring alternative management policies at levels ranging from individual organizations to national strategy. Often, the organizations of interest are best modeled using multilevel representations. The interactions with simulators of such complex systems almost always involve groups of people rather than individuals, often with different stakeholders in conflict about priorities and courses of action.

Multilevel Modeling

To develop policy flight simulators, we need to computationally model the functioning of the complex system of interest to enable decision makers, as well as other significant stakeholders, to explore the possibilities and implications of transforming these enterprise systems in fundamental ways. The goal is to create organizational simulations that will serve as policy flight simulators for interactive exploration by teams of often disparate stakeholders who have inherent conflicts, but need and desire an agreed-upon way forward (Rouse & Boff, 2005).

Consider the architecture of public–private enterprises shown in Figure 8.7 (Rouse, 2009; Rouse & Cortese, 2010; Grossman et al., 2011). The efficiencies that can be gained at the lowest level (work practices) are limited by nature of the next level (delivery operations). Work can only be accomplished within the capacities provided by available processes. Further, delivery organized around processes tends to result in much more efficient work practices than for functionally organized business operations.

However, the efficiencies that can be gained from improved operations are limited by the nature of the level above, that is, system structure. Functional operations are often driven by organizations structured around these functions, for example, manufacturing and service. Each of these organizations may be a different business with independent economic objectives. This may significantly hinder process-oriented thinking.

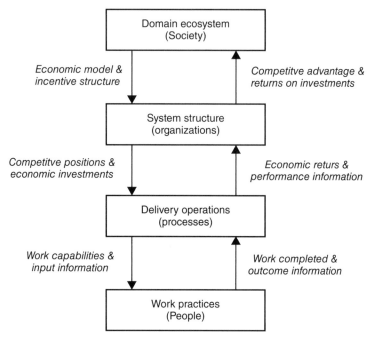

Figure 8.7   Architecture of Public–Private Enterprises

Of course, potential efficiencies in system structure are limited by the ecosystem in which these organizations operate. Market maturity, economic conditions, and government regulations will affect the capacities (processes) that businesses (organizations) are willing to invest in to enable work practices (people), whether these people be employees, customers, or constituencies in general. Economic considerations play a major role at this level (Rouse, 2010a, b).

These organizational realities have long been recognized by researchers in socio-technical systems (Emery & Trist, 1973), as well as work design and system ergonomics (Hendrick & Kleiner, 2001). The contribution of the concept of policy flight simulators is the enablement of computational explorations of these realities, especially by stakeholders without deep disciplinary expertise in these phenomena.

Example – Employee Prevention and Wellness

Developing multilevel models of large-scale public–private enterprises is a challenge in itself. Getting decision makers and other stakeholders to employ these models to inform their discussions and decisions is yet a greater challenge. We have found that interactive simulation models can provide the means to meeting this challenge. The decision makers with whom we have worked have found that the phrase "policy flight simulator" makes sense to them.

Multilevel simulations can provide the means to explore a wide range of possibilities, thereby enabling the early discarding of bad ideas and refinement of good ones. This enables the previously mentioned driving the future before writing the check. One would never develop and deploy an airplane without first simulating its behavior and performance. However, this happens all too often in enterprise systems in terms of policies, strategies, plans, and management practices that are rolled out with little, if any, consideration of higher-order and unintended consequences.

One of our policy flight simulators focused on the Predictive Health Institute (PHI), a joint initiative of Emory University and Georgia Institute of Technology (Brigham, 2010; Rask, Brigham & Johns, 2011). PHI is a health-focused facility that counsels essentially healthy people on diet, weight, activity, and stress management. The multilevel model focused on the roughly 700 people in PHI's cohort and their risks of type 2 diabetes (DM) and coronary heart disease. We calculated every person's risk of each disease using well-accepted risk models based on national data sets (Wilson et al., 1998, 2007), using PHI's initial individual assessments of blood pressure, fasting glucose level, and so on for each participant. Subsequent assessment data were used to estimate annual risk changes as a function of initial risks of each disease.

The four-level model of Figure 8.7 was implemented as a multilevel simulation. Separate displays were created to portray the operation of each level of the model. Runs of the multilevel simulation were set up using the dashboard in Figure 8.3. The top section of the dashboard allows decision makers to test different combinations of policies from the perspective of HR. For instance, this level determines the allocation of payments to PHI based on a hybrid capitated or pay-for-outcome formula. It also involves choices of parameters such as projected healthcare inflation rate, general economy inflation rate, and discount rate that affect the economic valuation of the outcomes of PHI's prevention and wellness program. One of the greatest concerns of HR is achieving a satisfactory ROI on any investments in prevention and wellness.

Note that the notion of payer is different in the United States than in most other developed countries. The payer for the majority of people covered by health insurance in the United States is their employer. Payment is usually managed by the HR function in companies, often with assistance from the private insurance company that administers the company's plan. Companies are almost always concerned that healthcare expenditures are well managed and provide "returns" in terms of the well-being of employees and their families, as well as the performance of employees in their jobs (Rouse, 2010b).

The concerns of PHI are represented in the lower section of the dashboard. These concerns include the organization's economic sustainability – their revenue must be equal to or greater than their costs. To achieve sustainability, PHI must appropriately design its operational processes and rules. Two issues are central. What risk levels should be used to stratify the participant population? What assessment and coaching processes should be employed for each strata of the population? Other considerations at this level include the growth rate of the participant population, the age ranges targeted for growth, and the program duration before participants are moved to "maintenance."

Decision makers can also decide what data sources to employ to parameterize the models – either data from the American Diabetes Association and American Heart Association, or data specific to Emory employees. Decision makers can choose to only count savings until age 65 or also project postretirement savings.

This policy flight simulator was used to explore two scenarios: (i) capitated payment for services and (ii) payment for outcomes. Hybrids of these scenarios were also investigated (Park et al., 2012). The goal was to understand the influence of capitation and pay-for-outcome levels on economic outcomes for both payer (HR) and provider (PHI).

Since PHI delivers the same service to all volunteers, a pure capitated payment is essentially a fee for service. PHI can be very profitable if the capitated payment is sufficiently large. On the other hand, PHI does only modestly well by comparison under a payment for outcomes system, in large part because its population is not prescreened for people at risk.

Emory HR's results are virtually opposite, although it can still do relatively well under the right blend of capitation and pay for outcome. The aggregate results for Emory as a whole (i.e., PHI plus HR) provide a surrogate for "society" and its overall gain under various healthcare payment systems. Here the results are far less intuitive and, in fact, a typical negotiation that finds middle ground, for example, by "splitting the difference" between HR and PHI, would not achieve anything close to the maximum potential overall societal gain.

In other words, when we compromise between the returns to HR and PHI, the aggregate returns to Emory are minimized. The best economic results are achieved when *either* PHI's profit is maximized or Emory HR's ROI is maximized. There are a variety of reasons why one might choose either extreme. However, another possibility emerged from discussions while using the policy flight simulator.

HR could maximize its ROI while providing PHI a very lean budget. At the end of each year, HR could then provide PHI with a bonus for the actual savings experienced that year. This could be determined by comparing the projected costs for the people in the program to their actual costs of health care, absenteeism, and presenteeism. In this way, HR would be sharing actual savings rather than projected savings. The annual bonuses would free PHI of the fear of not being sustainable, although PHI would need to substantially reorganize its delivery system to stratify participants by risk levels and tailor processes to each stratum.

This policy flight simulator was used to explore a wide range of other issues such as the best levels of participant risk stratification and the impacts of inflation and discount rates. The key insight for PHI management was that they needed to redesign their processes and decision rules if they were going to provide a good return to HR and stay in business. They learned how best to redesign their offerings using the policy flight simulator. Now they are getting ready for flight tests.

People's Use of Simulators

There are eight tasks associated with creating and using policy flight simulators:

- Agreeing on objectives – the questions – for which the simulator will be constructed.
- Formulating the multilevel model – the engine for the simulator – including alternative representations and approaches to parameterization.
- Designing a human–computer interface that includes rich visualizations and associated controls for specifying scenarios.
- Iteratively developing, testing, and debugging, including identifying faulty thinking in formulating the model.
- Interactively exploring the impacts of ranges of parameters and consequences of various scenarios.
- Agreeing on rules for eliminating solutions that do not make sense for one or more stakeholders.
- Defining the parameter surfaces of interest and "production" runs to map these surfaces.
- Agreeing on feasible solutions and the relative merits and benefits of each feasible solution.

The discussions associated with performing the aforementioned tasks tend to be quite rich. Initial interactions focus on agreeing on objectives, which includes output measures of interest, including units of measure. This often unearths differing perspectives among stakeholders.

Attention then moves to discussions of the phenomena affecting the measures of interest, including relationships among phenomena. Component models are needed for these phenomena and agreeing on suitable vetted, and hopefully off-the-shelf, models occurs at this time. Also of great importance are uncertainties associated with these phenomena, including both structural and parametric uncertainties.

As computational versions of models are developed and demonstrated, discussions center on the extent to which model responses are aligned with expectations. The overall goal is to computationally redesign the enterprise. However, the initial goal is usually to replicate the existing organization to see if the model predicts the results actually being currently achieved.

Once attention shifts to redesign, discussion inevitably shifts to the question of how to validate the model's predictions. As these predictions inherently concern organizational systems that do not yet exist, validation is limited to discussing the believability of the insights emerging from debates about the nature and causes of model outputs. In some cases, deficiencies of the models will be uncovered, but occasionally unexpected higher-order and unintended consequences make complete sense and become issues of serious discussion.

Model-based policy flight simulators are often used to explore a wide range of ideas. It is quite common for one or more stakeholders to have bright ideas that have substantially negative consequences. People typically tee up many alternative organizational designs, interactively explore their consequences, and develop criteria for the goodness of an idea. A common criterion is that no major stakeholder can lose in a substantial way. For the Emory simulator, this rule pared the feasible set from hundreds of thousands of configurations to a few hundred.

Quite often, people discover the key variables most affecting the measures of primary interest. They then can use the simulator in a "production mode," without the graphical user interface, to rapidly simulate ranges of variables to produce surface plots. The simulator runs to create these plots are done without the user interface of Figure 8.3.

Discussions of such surface plots, as well as other results, provide the basis for agreeing on pilot tests of apparently good ideas. Such tests are used to empirically confirm the simulator's predictions, much as flight tests are used to confirm that an aircraft's performance is similar to that predicted when the plane was designed "in silico."

Policy flight simulators serve as boundary spanning mechanisms, across domains, disciplines, and beyond initial problem formulations, which are all too often more tightly bounded than warranted. Such boundary spanning results in arguments among stakeholders being externalized. The alternative perspectives are represented by the assumptions underlying and the elements that compose the graphically depicted model projected on the large screen. The debate then focuses on the screen rather than being an argument between two or more people across a table.

The observations in this section are well aligned with Rouse's (1998) findings concerning what teams seek from computer-based tools for planning and design:

- Teams want a clear and straightforward process to guide their decisions and discussions, with a clear mandate to depart from this process whenever they choose.
- Teams want capture of information compiled, decisions made, and linkages between these inputs and outputs so that they can communicate and justify their decisions, as well as reconstruct decision processes.
- Teams want computer-aided facilitation of group processes via management of the nominal decision-making process using computer-based tools and large screen displays.
- Teams want tools that digest the information that they input, see patterns or trends, and then provide advice or guidance that the group perceives they would not have thought of without the tools.

Policy flight simulators do not yet fully satisfy all these objectives, but they are headed in this direction.

It is useful to note that the process outlined in this section is inherently a participatory design process (Schuler & Namioka, 1993). This human-centered process considers and balances all stakeholders' concerns, values, and perceptions (Rouse, 2007). The result is a better solution and, just as important, an acceptable solution.

CONCLUSIONS

This chapter first briefly addressed human vision as a phenomenon, primarily to recognize the topic as important but also to move beyond the science of vision to the design of visualizations. We next reviewed the basics of visualization to provide grounding for the subsequent sections. The next section addressed the purposes of visualization, the object of design being the fulfillment of purposes. A visualization design methodology was then presented and illustrated with an example from helicopter maintenance. Visualization tools were then briefly reviewed. Various case studies from Stevens Institute's *Immersion Lab* were discussed. The notion of policy flight simulators was introduced and elaborated. Finally, results were presented from an extensive study of what users want from visualizations and supporting computational tools. We next consider how to populate elements of visualizations with underlying computational models.

REFERENCES

Basole, R.C. (2009). Visualization of interfirm relations in a converging mobile ecosystem. *Journal of Information Technology*, 24 (2), 144–159.

Basole, R.C., Bodner, D.A., & Rouse, W.B. (2013). Healthcare management through organizational simulation. *Decision Support Systems*, 55 (2), 552–563.

Biederman, I. (1987). Recognition-by-components: A theory of human image understanding. *Psychological Review*, 94 (2), 115–147.

Blumberg, A. (2013). Demonstration of Hoboken Flooding. Hoboken, NJ: Davidson Laboratory, Stevens Institute of Technology.

Boardman, J.T., & Cole, A.J. (1996). Integrated process improvement in design and manufacture using a systems approach *IEE Proceedings on Control Theory and Applications*, 143 (2) 171–185.

Brigham, K.L. (2010). Predictive health: The imminent revolution in healthcare. *Journal of the American Geriatrics Society*, 58 (S2), S298–S302.

Burton, R.M. (2003). Computational laboratories for organization science: Questions, validity and docking. *Computational and Mathematical Organization Theory*, 9, 91–108.

Burton, R.M., & Obel, B. (2011). Computational modeling for what-is. What-might-be and what-should-be studies – and triangulation. *Organization Science,* 22 (5), 1195–1202.

Carley, K. (2002). Computational organization science: A new frontier. *Proceedings of the National Academy of Sciences of the United States of America*, 99 (3), 7257–7262.

Chen, C. (2004). Searching for intellectual turning points: Progressive knowledge domain visualization. *Proceedings of the National Academy of Sciences of the United States of America*, 101 (1), 5303–5310.

Chen, C. (2005). Top 10 unsolved information visualization problems. *IEEE Computer Graphics and Applications*, 25 (4), 12–16.

Emery, F. & Trist, E. (1973). Toward a Social Ecology. London: Plenum Press,

Frey, P.R., Rouse, W.B., & Garris, R. D. (1992). Big graphics and little screens: Designing graphical displays for maintenance tasks. *IEEE Transactions on Systems, Man, and Cybernetics*, 22 (1), 10–20.

Frey P.R., Rouse, W.B., & Garris, R.D. (1993). Big graphics and little screens: Model-based design of large-scale information displays, in W.B. Rouse, Ed., Human/Technology Interaction in Complex Systems (Vol. 6, pp. 1–57). Greenwich, CT: JAI Press.

Grossman, C., Goolsby, W.A., Olsen, L., & McGinnis, J. M. (2011). Engineering the Learning Healthcare System. Washington, DC: National Academy Press.

Hendrick, H.W., & Kleiner, B.M. (2001). Macroergonomics: An Introduction to Work System Design. Santa Monica, CA: Human Factors and Ergonomics Society.

Howard, R.A., & Matheson, J.E. (1981). Influence diagrams, in R.A. Howard & J.E. Matheson, Eds., Readings on the Principles and Applications of Decision Analysis, Vol. I (pp. 721–762). Menlo Park, CA: Strategic Decisions Group.

Klein, G. (2003). Intuition at Work: Why Developing Your Gut Instincts Will Make You Better at What You Do. New York: Doubleday.

Marr, D. (1982). Vision: A Computational Investigation into the Human Representation and Processing of Visual Information. San Francisco: Freeman.

Minsky, M. (1988). The Society of Mind. New York: Simon & Schuster.

Minsky, M., & Papert, S. (1969). Perceptrons. Cambridge, MA: MIT Press.

Moggridge, B. (2007). Designing Interactions. Cambridge, MA: MIT Press.

Norman, D.A. (1988). The Psychology of Everyday Things. New York: Basic Books.

Park, H., Clear, T., Rouse, W.B., Basole, R.C., Braunstein, M.L., Brigham, K.L., & Cunningham, L. (2012). Multi-level simulations of health delivery systems. A prospective tool for policy, strategy, planning and management. *Journal of Service Science*, 4 (3), 253–268.

Prietula, M., Carley, K., & Gasser, L. (Eds.). (1998). Simulating Organizations: Computational Models of Institutions and Groups. Menlo Park, CA: AAAI Press.

Rask, K.J., Brigham, K.L., & Johns, M.M.E. (2011). Integrating comparative effectiveness research programs into predictive health: A unique role for academic health centers. *Academic Medicine*, 86 (6), 1–6.

Rasmussen, J. (1983). Skills, rules, and knowledge; Signals, signs, and symbols, and other distinction in human performance models. *IEEE Transactions on Systems, Man, and Cybernetics*, 14 (3), 257–266.

Rasmussen, J. (1986). Information Processing and Human-Machine Interaction. New York: North-Holland.

Rouse, W.B. (1983). Models of human problem solving: Detection, diagnosis, and compensation for system failures. *Automatica*, 19(6), 613–625.

Rouse, W.B. (1998). Computer support of collaborative planning. *Journal of the American Society for Information Science*, 49 (9), 832–839.

Rouse, W.B. (2007). People and Organizations: Explorations of Human-Centered Design. New York: Wiley.

Rouse, W.B. (2009). Engineering perspectives on healthcare delivery: Can we afford technological innovation in healthcare? *Journal of Systems Research and Behavioral Science*, 26, 1–10.

Rouse, W.B. (2010a). Impacts of healthcare price controls: Potential unintended consequences of firms' responses to price policies, *IEEE Systems Journal*, 4 (1), 34–38.

Rouse, W.B. (Ed.). (2010b). The Economics of Human Systems Integration: Valuation of Investments in People's Training and Education, Safety and Health, and Work Productivity. New York: Wiley.

Rouse, W.B., & Boff, K.R. (Eds.). (2005). Organizational Simulation: From Modeling and Simulation to Games and Entertainment. New York: Wiley.

Rouse, W.B., & Cortese, D.A. (Eds.). (2010). Engineering the System of Healthcare Delivery. Amsterdam: IOS Press.

Rouse, W.B., & Howard, C.W. (1995). Supporting market-driven change, in D. Burnstein, Ed., The Digital MBA (pp. 159–184). New York: Osborne McGraw-Hill.

Rouse, W.B., & Serban, N. (2014). Understanding and Managing the Complexity of Healthcare. Cambridge, MA: MIT Press.

Rouse, W.B., Howard, C.W., Carns, W.E., & Prendergast, E.J. (2000). Technology investment advisor: An options-based approach to technology strategy. *Information Knowledge Systems Management*, 2 (1), 63–81.

Schuler, D. & Namioka, A. (1993). Participatory Design: Principles and Practices. Hillsdale, NJ: Erlbaum.

Tufte, E.R. (1983). The Visual Display of Quantitative Information. Cheshire, CT: Graphics Press.

Tufte, E.R. (1997). Visual Explanations: Images and Quantities, Evidence and Narrative. Cheshire, CT: Graphics Press.

Ware, C. (2012). Information Visualization (3rd ed.). Burlington, MA: Morgan Kaufmann.

Wilson, P.W., D'Agostino, R.B., Levy, D., Belanger, A.M., Silbershatz, H., & Kannel, W.B. (1998). Prediction of Coronary Heart Disease Using Risk Factor Categories. Framingham, MA: Framingham Heart Study, National Heart, Lung and Blood Institute.

Wilson, P.W., Meigs, J.B., Sullivan, L., Fox, C.S., Nathan, D.M., & D'Agostino, R.B. (2007). Prediction of incident diabetes mellitus in middle aged adults: The Framingham offspring study. *Archives of Internal Medicine*, 167, 1068–1074.

Yu, Z. (2014). *Computational methods for supporting enterprise transformation* [PhD dissertation], School of Systems and Enterprises, Stevens Institute of Technology.

9

COMPUTATIONAL METHODS AND TOOLS

INTRODUCTION

It is useful to begin by discussing where we are methodologically and, more generally, where we are in the overall problem-solving process. At this point, we are ready to begin Step 5 of our overall methodology. Thus, we have completed at least an initial pass of Steps 1–4:

- *Step 1*: Decide on the central questions of interest;
- *Step 2*: Define key phenomena underlying these questions;
- *Step 3*: Develop one or more visualizations of relationships among phenomena;
- *Step 4*: Determine key trade-offs that appear to warrant deeper exploration.

Steps 5–7 are the focus of this chapter. Progress on these steps is likely to result in revisiting Steps 1–4 as, for example, digging deeper results in identifying one or more important phenomena not previously considered. In general, however, we are ready to consider how we might address the trade-offs resulting from Step 4 using mathematical and computational methods and tools.

Modeling and Visualization of Complex Systems and Enterprises:
Explorations of Physical, Human, Economic, and Social Phenomena, First Edition. William B. Rouse.
© 2015 John Wiley & Sons, Inc. Published 2015 by John Wiley & Sons, Inc.

Table 9.1 summarizes the phenomena associated with the six archetypal problems discussed throughout this book and addressed in detail in Chapters 4–7. For each instance of a phenomenon, potentially useful modeling paradigms are listed. This chapter elaborates these paradigms, while also providing a brief overview of software tools available to support use of these approaches.

It is important to note that these modeling paradigms are not required for pursing the archetypal problems. I have seen useful, albeit usually quite simple, models developed in Microsoft Excel, as well as very elaborate models developed from first principles and represented in many thousands of lines of software code. Consideration of the modeling paradigms listed will, however, provide well-founded ways to approach the archetypal problems and, of course, many others.

This chapter proceeds as follows. Modeling paradigms potentially useful for addressing the phenomena in Table 9.1 are next discussed, including dynamic systems theory, control theory, estimation theory, queuing theory, network theory, decision theory, problem-solving theory, and finance theory. A multilevel modeling framework is used to illustrate how the different modeling paradigms can be employed to represent different levels of abstraction and aggregation of an overall problem. The next consideration is moving from representation to computation. This includes discussion of model composition and issues of entangled states and consistency of assumptions. This chapter also provides a brief overview of software tools available to support use of these computational approaches.

MODELING PARADIGMS

In this section, we discuss the following modeling paradigms:

- *Dynamic systems theory*. Ordinary differential equations and partial differential equations;
- *Control theory*. Classical feedback control and optimal control theory;
- *Estimation theory*. Statistical estimation, stochastic estimation, and pattern recognition;
- *Queuing theory*. Markov processes and waiting line theory;
- *Network theory*. Graph theory, agent-based models, and routing problems;
- *Decision theory*. Utility theory, Bayes theory, game theory, social choice theory;

TABLE 9.1 Archetypal Phenomena and Modeling Paradigms

Category	Phenomenon	Modeling Paradigm
Physical, natural	Flow of water	Dynamic systems theory
Physical, natural	Disease incidence/progression	Statistical models, Markov processes
Physical, natural	Cell growth and death	Network theory, biochemistry
Physical, natural	Biological signaling	Network theory, biochemistry
Physical, designed	Flow of parts	Network theory, queuing theory
Physical, designed	Assembly of parts	Network theory, queuing theory
Physical, designed	Flow of demands	Network theory, queuing theory
Physical, designed	Traffic congestion	Network theory, dynamic systems theory
Physical, designed	Vehicle flow	Agent-based models
Physical, designed	Infrastructure response	Dynamic systems theory, network theory
Human, individual	Diagnosis decisions	Pattern recognition, problem solving
Human, individual	Selection decisions	Decision theory
Human, individual	Control performance	Dynamic systems theory, control theory
Human, individual	Perceptions and expectations	Pattern recognition, Bayes theory
Human, team/group	Group decision making	Decision theory, social choice theory
Economic, micro	Investment decision making	Decision theory, discounted cash flow
Economic, micro	Operational decision making	Network theory, optimization
Economic, micro	Risk management	Decision theory, Bayes theory
Economic, micro	Dynamics of competition	Game theory, differential equations
Economic, macro	Dynamics of demand and supply	Dynamic systems theory, optimization
Economic, macro	Prices, costs, and payment	Discounted cash flow, optimization
Social, information sharing	Social networks	Network theory, agent-based models
Social, organizations	Domain social system	Network theory, decision theory
Social, values/norms	Domain values and norms	Network theory, decision theory

- *Problem solving theory.* Deductive and inductive reasoning, recognition-primed problem solving;
- *Finance theory.* Discounted cash flow.

Table 9.2 summarizes the key assumptions underlying each modeling paradigm and the central phenomena predicted using these paradigms. Note that we need to differentiate predicted phenomena such as stability and equilibrium from the physical, human, economic, and social phenomena discussed in Chapters 4–7. Most of predicted phenomena in the right column of Table 9.2 could be applied to a wide variety of problems in many domains. In contrast, many of the phenomena in the center column of Table 9.1 are rather domain specific.

Dynamic Systems Theory

A formal theory of dynamic systems dates back at least to Isaac Newton in the 17th century. This led to the development of differential equations that enable relating some function of one or more variables with its derivative or infinitesimal rate of change. Differential equations developed from calculus, which was independently invented by Newton and Gottfried Liebniz.

Differential equation representations of dynamic systems are typically derived from "first principle" descriptions of the phenomena associated with the system. Typical assumptions include Newton's laws, conservation of mass, and continuity of transport. Solutions of the resulting differential equations are used to predict the magnitude of a system's response, the time required to fully respond, and the stability of the response in the sense that bounded inputs yield bounded outputs.

An ordinary differential equation includes a function of one independent variable and its derivatives. Partial differential equations include more than one independent variable and derivatives. The equations for urban oceanography discussed in Chapter 4 are partial differential equations. Multiple independent variables are needed to represent the spatial nature of oceans.

Differential equations are often stated as difference equations, particularly when digital computers are used to solve the equations and possibly control the dynamic systems they represent. Equation (9.1) shows a basic difference equation for a linear system.

$$\underline{X}(t+1) = \underline{\Phi}\,\underline{X}(t) + \underline{\Gamma}\,\underline{W}(t) \qquad (9.1)$$

where $\underline{X}$ is an N dimensional state vector, $\underline{\Phi}$ is an $N \times N$ state transition matrix, $\underline{\Gamma}$ is an $N \times P$ disturbance transition matrix, and $\underline{W}$ is an input vector, typically termed a disturbance. This equation is solved by first specifying initial conditions, that is, $\underline{W}(1)$. Then, given $\underline{W}(t)$ for $t > 0$, one can compute $\underline{X}(t)$ for $t > 1$.

TABLE 9.2 Modeling Paradigms, Typical Assumptions, and Phenomena Predicted

Modeling Paradigm	Typical Assumptions	Phenomena Predicted
Dynamic systems theory	• Newton's laws • Conservation of mass • Continuity of transport	• Response magnitude • Response time • Stability of response
Control theory	• Known transfer function of state transition matrix • Stationary, Gaussian stochastic processes • Given objective function of errors, control effort	• Response time • Stability of response • Control errors • Observability • Controllability
Estimation theory	• Known dynamics of process • Known ergodic (stationary) stochastic process • Additive noise inputs	• State estimates – filtering, smoothing, prediction • Estimation errors
Queuing theory	• Known arrival and service processes • Future state only depends on current state • Given service protocol, for example, First Come, First Served, priority	• Number and time in queue • Number and time in system • Probability of balk or renege
Network theory	• Discrete entities, for example, agents • Decision rules of entities, for example, agents • Typically binary relationships • Relationships only via arcs or edges	• Shortest distance between any two locations (nodes) • Shortest time between any two locations (nodes) • Propagation of sentiment among actors

(*continued*)

TABLE 9.2   (*Continued*)

Modeling Paradigm	Typical Assumptions	Phenomena Predicted
Decision theory	• Known utility functions • Comparable utility metrics • Known payoff matrix • Given voting rules	• Choice selected • Game equilibrium • Election results • Impacts of incentives
Problem-solving theory	• Known human mental model • Known information utilization • Known repertoire of patterns • Known troubleshooting rules	• Time until problem solved • Steps until problem solved • Problem-solving errors
Finance theory	• Projected investments • Projected operating costs • Projected revenues and costs	• Net present value • Net option value • Net capital at risk

Note that the "state" of a system is defined as the set variables such that knowledge of their values, in addition to knowledge of system inputs, enables prediction of future system outputs. If one is missing important variables from the definition of state, predictions of future states will be erroneous.

Figure 9.1 portrays this basic dynamic system, as well as a system under control, and a system that includes state estimation needed to compensate for stochastic disturbances. The latter two cases are discussed in the next two subsections.

Control Theory

It is often desirable to control the state of the system to achieve desirable values. This requires that we expand the difference equation to include additional terms.

$$\underline{X}(t+1) = \underline{\Phi}\,\underline{X}(t) + \underline{\Gamma}\,\underline{W}(t) + \underline{\Psi}\,\underline{U}(t) \tag{9.2}$$

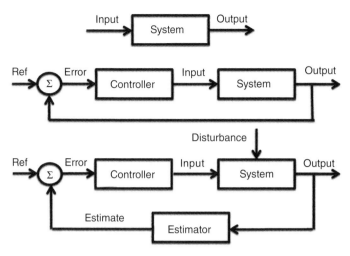

Figure 9.1   Basic Dynamic System, with Control, and with Estimator

$$\underline{Z}(t + 1) = \underline{H}\,\underline{X}(t + 1) + \underline{V}(t + 1) \tag{9.3}$$

where $\underline{\Psi}$ is an $N \times R$ input transition matrix, $\underline{H}$ is an $M \times N$ observation transition matrix, and $\underline{V}$ represents potential observation noise.

Maxwell formalized control theory beginning in the mid-1800s for machine governors. Control was essential to initial manned aircraft, including the first demonstration by the Wright Brothers. In World War II, control theory was central to fire control systems and guidance systems.

Modern control theory focuses on time-domain state-space representations with variables expressed as vectors and, for linear systems, differential equations replaced by matrix algebra. Optimal control theory provides a means for deriving control policies. Bellman (1967) and Pontryagin (1962) developed this theory. A special case of optimal control is the linear quadratic Gaussian (LQG) formulation where the differential or difference equations are linear, the function to be optimized is a quadratic function of state variations and control energy, and all disturbances are additive white Gaussian noise (Athans, 1971).

Typical assumptions when employing control theory formulations include known transfer functions or state transition matrices; stationary, Gaussian stochastic processes; and given objective functions of state variations and control energy. Predictions sought when solving these models includes response time, stability of response, and control errors.

Observability and controllability are two very important concepts in modern control theory (Kalman, 1961). Observability concerns the extent to which the state of a system can be inferred by knowledge of its external outputs. Observability can be assessed using $\underline{\Phi}$ and $\underline{H}$. Controllability denotes the ability of a control input to move the state of a system from any initial state to any other final state in a finite time interval. Controllability can be assessed using $\underline{\Phi}$ and $\underline{\Psi}$. Being able to observe, if only by inference, the full state of a system and control the full state to any desired values are central to optimal control theory.

Both classic and modern control theory have often been used to model human control of aircraft, ships, automobile, process plants, and so on. Rasmussen (1986), Rasmusses et al. (1994), Rouse (1980, 2007), Sheridan and Ferrell (1974), and Sheridan (1992) review many of these applications of control theory. When humans' performance must comply with the constraints of an engineered system operating within a demanding, yet structured, environment, control theory models are often quite successful.

Estimation Theory

If the stochastic components of W and V in equations (9.2) and (9.3) are of sufficient magnitude, then the control problem includes an estimation problem. This is a stochastic estimation problem unlike parameter estimation problems from statistics. A Kalman filter (1960) can be used to obtain an estimate of system state, denoted by $\hat{}$, using equation (9.4).

$$\hat{\underline{X}}(t + 1|t + 1) = \underline{\Phi}\,\hat{\underline{X}}(t|t) + \underline{K}(t + 1)[\underline{Z}(t + 1) - \underline{H}\,\underline{\Phi}\,\hat{\underline{X}}(t|t)] \qquad (9.4)$$

where $\underline{K}$ is an $N \times M$ Kalman filter gain matrix. Thus, the filtered estimate is a linear weighted function of the projected system state and the difference between the actual and predicted observed system output. This difference, termed the residual, should be a zero-mean Gaussian process if the filter is working correctly.

Typical assumptions associated with stochastic estimation are known dynamics of process, known ergodic (stationary) stochastic process, and additive noise inputs. Thus, $\underline{\Phi}$, $\underline{\Gamma}$, $\underline{\Psi}$, and $\underline{H}$ are assumed known, as are the distributional characteristics of $\underline{W}$ and $\underline{V}$. Given these assumptions, one can then compute state estimates – for filtering, smoothing, or prediction – and project estimation errors.

Stochastic estimation theory has been used to model human performance in tasks ranging from failure detection to air traffic control (Rouse, 1983, 2007). In some of these applications, it was unreasonable to assume that humans knew the system's dynamics, that is, $\underline{\Phi}$. Hence, the estimation problem had to be expanded to include ongoing estimation of the parameters of $\underline{\Phi}$. This is termed a system identification problem. An identifier with fading memory (i.e., older observations are weighted less than more recent observations) enabled formulating a model that predicted human performance quite well.

Queuing Theory

The previous three sections dealt with the dynamics of a system's response and how to control the response. The focus was on errors between desired and actual response traded off against the energy required to minimize these errors. Queuing theory focuses on the time spent by entities waiting for service and in service while in a service system.

Erlang first formalized queuing theory in 1909, although the notion of queuing systems did not emerge until after World War II. Morse (1958) published the first textbook on queuing. The ideas have been applied in many domains ranging from telecommunications to traffic engineering and design of factories, retail stores, and hospitals.

Figure 9.2 depicts a basic queuing system, including its Markov state transition diagram. The state of a queuing system equals the number of entities, n, in the system. The probability P of the system being in state n is given by

$$P_n = \left(1 - \frac{\lambda}{\mu}\right)\left(\frac{\lambda}{\mu}\right)^n \qquad (9.5)$$

where λ is the arrival rate of a Poisson process and μ is the service rate of an exponential distribution, and there is assumed to be a single server.

The average number of entities in the system L_S and the average number waiting in the queue are L_Q given by

$$L_S = \frac{\lambda}{\mu - \lambda} \qquad (9.6)$$

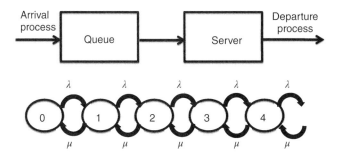

Figure 9.2   Basic Queuing System

$$L_Q = \frac{\lambda^2}{\mu(\mu - \lambda)} \tag{9.7}$$

The average time in the system W_S and average time spent waiting in the queue W_Q are given by

$$W_S = \frac{1}{\mu - \lambda} \tag{9.8}$$

$$W_Q = \frac{\lambda}{\mu(\mu - \lambda)} \tag{9.9}$$

Typical assumptions employed in developing a queuing theory model include known arrival and service processes (e.g., Poisson, exponential), future state only depending on current state (i.e., Markov processes), and given service protocol (e.g., First Come, First Served, priority). Responses predicted include number of entities and time they spend in queue and in the system (i.e., waiting plus in service), and the probabilities of balking (i.e., refusing to enter the system) or reneging (i.e., leaving without being served).

There have been an enormous range of applications of queuing theory, often via simulation for complex situations where the assumptions needed to enable analytical solutions are unlikely to be warranted. As discussed in Chapter 5, a particularly interesting application involves predicting how humans handle multitask decision making (Rouse, 2007). How do they prioritize multiple queues of demands and allocate their attention? For example, among the many things competing for an aircraft pilot's attention, can we predict what they will get to and when? Such predictions have been used to determine when the pilot needs help – typically at a point where he or she is too busy to ask for help.

Queuing theory helped us to understand how to design such an adaptive aiding system. It helped us to determine when an artificially intelligent "pilot's associate" should offer help. There are, of course, other approaches to modeling how humans decide what to do next, for example, Simon (1978). However, when a strong formalism like queuing theory is applicable, the analytic tools available can be quite powerful.

Network Theory

The origination of network or graph theory is usually attributed to Euler and his 1736 paper on the bridges of Konigsberg in Prussia. The problem addressed was finding a route through the city that crossed each of the seven bridges just once. The term "graph" was introduced 150 years later and the first textbook was published 50 years after that.

Graphs are composed of nodes or vertices and arcs or edges that connect them. Arcs may be directed, hence unidirectional, or undirected. In chemistry, nodes can represent atoms while arcs denote bonds. Graphs can be used to represent language in terms of relationships between words. In behavioral and social science, nodes can represent people while arcs denote relationships.

Typical assumptions underlying network models include the discrete nature of entities, for example, people, the decision rules, if any, of these entities, limiting relationships to binary, and only representing relationships via arcs or edges. The resulting network depictions are usually context free, although the model is usually constructed to address questions within particular contexts.

Network models have long been used to address questions such as the shortest distance between any two locations (nodes) or the shortest time between any two locations (nodes). For models where nodes represent people, questions of interest include propagation of sentiment among voters or communications among citizens during emergencies. As discussed in Chapter 7, Batty (2013) has recently used network theory to understand the geographic and economic growth of cities.

Figure 9.3 represents a service network. Requests for service enter via nodes 1, 4, or 7, and exit via nodes 15, 19, or 20. Requests may be in the form of customers, patients, or laptops to be repaired. Routing of requests through the network depends on the particular services needed as well as the workload of alternative nodes that provide the needed services.

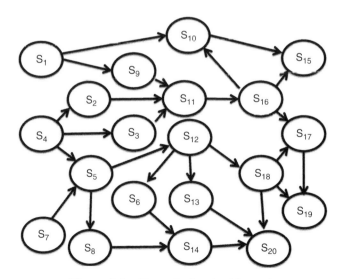

Figure 9.3   Example Service Network

We have studied these types of networks in the context of statewide information services networks (Rouse, 2007) as well as healthcare delivery networks (Rouse & Serban, 2014). In order to compute the performance of these networks, we first define a matrix of transitions probabilities, $\underline{P}$, with element p_{ij} denoting the probability of transitioning from node i to node j. Assume that there are K sources of service requests and J classes of requests. We define λ_{jk} for $j = 1, 2, \ldots, J$, and $k = 1, 2, \ldots, K$, as the average demand per unit time in class j from source k. This is termed type jk demand.

Type jk demand is routed through the network using a routing vector $\underline{R}_{jk}$, with elements r_{jkl}, for $l = 1, 2, \ldots, L_{jk}$, where the lth element of the routing vector is the designation of the service resource to which the request will be routed if it has not been satisfied by the first $l - 1$ elements of the routing vector. L_{jk} is the length of the route for type jk demand.

Given the average demands and routing vectors (or policies), we would like to determine λ_{ijkl}, which is the average type jk demand received by service resource i for $i = 1, 2, \ldots, I$ (upper case i) at routing stage l (lower case L). This is termed type $ijkl$ demand. For a request to become a type $ijkl$ request, it must be unsatisfied through its first $l - 1$ routing stages. Thus, to compute λ_{ijkl} we must determine the probability of satisfaction for a type jk request at each of its routing stages. The probabilities of interest are, therefore, conditional probabilities.

To determine these conditional probabilities, each service resource is modeled as subnetwork of M_i nodes with transition probability network $\underline{P}_{ijl}$. Subnodes receive requests for services from both the larger network and other nodes in the subnetwork. A bit of matrix algebra leads to the following relationship:

$$\underline{\Lambda}_{ijl} = (\underline{I} - \underline{P}'_{ijl})^{-1} \underline{\lambda}_{ijl} \tag{9.10}$$

where $\underline{\Lambda}$ denotes total demand and I is an $M_i \times M_i$ identify matrix.

We can compute the average demand on service resource i to be the vector

$$\underline{\Lambda}_i = \Sigma_j \Sigma_l \underline{\Lambda}_{ijl} \tag{9.11}$$

The average request processing time is defined by $\underline{w}_i$, an M_i vector of average times for requests to pass through the subnodes of service resource i. These times can be calculated using the queuing models discussed earlier or more elaborate queuing models. We can also compute the average queue length using similar models. Finally, we can similarly estimate $\underline{c}_i$, the vector of average processing costs to pass through the subnodes of service resource i.

Based on the routing vectors, we can determine the proportion of type jk requests that experience the performance of subnetwork i. We can then calculate p_{jk}, w_{jk}, and c_{jk}, the probability of success, average processing time, and average processing cost for type jk requests. These metrics, combined with λ_{jk}, enable summing across j and k to compute the probability P of the network satisfying a request, the average waiting time, W, from initial request to service receipts, and the average cost, C, per satisfied request.

P, W, and C are highly influenced by the routing vector $\underline{R}_{jk}$ and the length of the routes L_{jk} for each of the jk classes of demand. This queuing network model has been used to determine the optimal routes for each class of demand. A multiattribute utility function of P, W, and C was employed as the objective. Optimal routing led to priority being given to service resources with high probabilities of success and low waiting times and servicing costs. However, due to long waiting time and high servicing costs, the resources with highest probability of success were often later in the optimal routes. In this way, efficient and inexpensive resources provided services unless they were incapable of providing the requisite services. As a result, it was found that 10% of the state budget could be saved while preserving the service levels. Alternatively, this 10% could be invested to achieve significantly improved service levels.

Decision Theory

Decision theory emerged in the 18th century, building on innovations of three people. Daniel Bernoulli (1700–1782) was a pioneer in probability and statistics. Thomas Bayes (1702–1761) was a statistician renown for Bayes rule regarding the updating of conditional probabilities. Jeremy Bentham (1748–1832) was a philosopher best known for utilitarianism. The notion of game theory emerged during this period but was not formalized until much later.

The 1940s and 1950s saw enormous progress in decision theory. von Neumann and Morgenstern's *Theory of Games and Economic Behavior* (von Neumann & Morgenstern, 1944), Nash's theory of noncooperative games (Nash, 1950), and Arrow's social choice theory (Arrow, 1951) all appeared within a few years of each other. A couple of decades later, Keeney and Raiffa's classic *Decisions With Multiple Objectives* (1976) provided theory-based practical methods for decision analysis.

Typical assumptions underlying decision theory models include known utility functions and comparable utility metrics, all compatible with von Neumann and Morgenstern's axioms. For game theory, the payoff matrix is assumed known and availability of information is constrained. For social choice theory, voting rules are assumed known and followed.

Decision theory models are used to prescribe optimal decisions and predict choices selected, game equilibria, election results, and impacts of information and incentives. Equation (9.12) shows the standard utility function from Chapter 6. Equation (9.13) shows an example derived using Keeney and Raiffa's methodology. Equation (9.14) is an expected utility calculation.

$$U(\underline{X}) = U[u(x_1), u(x_2), \ldots, u(x_N)] \tag{9.12}$$

$$U(D, T) = 0.950\,u(D) + 0.455\,u(T) - 0.405\,u(D)\,u(T) \tag{9.13}$$

$$E[U(D, T)] = \sum_{i,j} p(D_i, T_j) \cup (D_i, T_j) \tag{9.14}$$

where D is price discount, T is delivery time, and $p\,(D_i, T_j)$ is the probability mass function for D and T.

In this example, the decision maker's preferences for trade-offs between price discounts and delivery times are represented. Discount is roughly twice as important as delivery time, but they are substitutable as indicated by the negative coefficient on the last term in equation (9.13). This model was central to developing a decision support system for procurement decisions (Rouse, 2007).

Figure 9.4 illustrates the modeling framework for new product planning as supported by the *Product Planning Advisor* (Rouse & Howard, 1995; Rouse, 2007), which was introduced in Chapter 8. The planning process begins by defining what the market wants in terms of different stakeholders, the attributes of concern to each stakeholder, and their utility functions for each attribute. An overall utility function for each stakeholder is defined by a weighted average of each the stakeholder's utility functions for each attribute. Similarly, an overall utility function across stakeholders is defined by a weighted average of each stakeholder's overall utility functions.

Attributes are grouped by validity, acceptability, and viability, three central constructs of human-centered design (Rouse, 1991, 2007). Validity is concerned with the extent to which a product or system solves the problem for which it is targeted. Acceptability focuses on the extent to which a product or system matches individual and organizational preferences for how a problem is solved. Viability concerns the benefits and costs of a solution, where costs include procurement, deployment, operation, maintenance of a solution, as well as the psychological costs of learning and changing how a problem is approached.

In the center of Figure 9.4, the relationships between functional descriptions of solutions and stakeholders attributes are specified using a seven-point

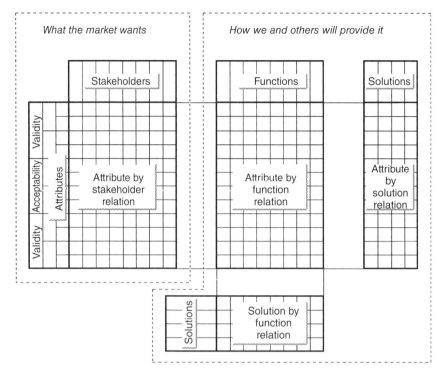

Figure 9.4 Modeling Framework Underlying *Product Planning Advisor*

scale from −3 to +3. Alternative solutions are described as groupings of functions on the right of Figure 9.4. Typically, both the user group's planned offerings are described as well as competitors' current and expected offerings. Many times users have focused on finding competitors' most likely responses to their own new offerings.

Product Planning Advisor has been used in many hundreds of new product planning efforts – several of these are discussed in Chapter 10. Users never limited themselves to simply identifying the solution with the highest expected utility. Instead, cross-disciplinary teams (e.g., engineering, manufacturing, finance, marketing, and sales) performed many "What if?" explorations, varying the relative importance attached to each attribute as well as the relative importance attached to each stakeholder. They also repeatedly used the "How to Improve" capabilities to determine how to change their offerings to be more competitive, as well as how their competitors were likely to do the same.

Problem-Solving Theory

Early studies of human problem solving began in the 1930s. Research in the 1960s and 1970s focused on simple, albeit novel, laboratory tasks, for example, Tower Of Hanoi. Allen Newell and Herbert Simon's *Human Problem Solving* (1972) is seen as a classic resulting from such studies. Their rule-based paradigm, which they termed production systems, became a standard in cognitive science.

More recently, that is, from the late 1970s until today, greater emphasis has been placed on the study of real problem solvers performing real-world tasks, e.g., (Rasmussen & Rouse, 1981). The research on detection and diagnosis of systems failures discussed in Chapter 5 and the research on displays for helicopter maintenance discussed in Chapter 8 are good examples of contextually rich studies of human problem solving.

Typical assumptions associated with models based on problem-solving theory include a specified human mental model of the problem domain and known information utilization behaviors, repertoire of symptom patterns, and troubleshooting rules. The phenomena usually predicted by such models include time until problem solved, steps until problem solved, and frequency of problem-solving errors.

We performed extensive studies of human diagnosis of systems failures in airplane, automobile and ship propulsion systems, power and process plants, and communications systems. A series of problem-solving models was developed and evaluated relative to actual troubleshooting behaviors and performance (Rouse, 1983, 2007). The last and most comprehensive model was a fuzzy rule-based model.

In Chapter 5, we discussed the notion of S-Rules and T-Rules, with S denoting symptomatic pattern-matching rules and T denoting topographic reasoning rules. S-Rules are very powerful – when they apply. T-Rules are less efficient but, when needed, will lead to successful problem solving. It is difficult to strictly order rules within either set or across sets.

This led to the idea that rules belong to fuzzy sets (Zadeh, 1965). Members of fuzzy sets have degrees of membership, rather than the binary membership associated with classic sets. We defined the fuzzy set of choosable rules to be the intersection of the fuzzy sets of recalled, applicable, useful, and simple rules. In fuzzy set theory, the intersection of sets is defined by the minimum of the membership functions of the component sets. Hence, the membership in the set of choosable rules is given by equation (9.15).

$$\mu_C = \text{Min}(\mu_R, \mu_A, \mu_U, \mu_S) \qquad (9.15)$$

where μ denotes membership functions for the fuzzy sets of choosable, recalled, applicable, useful, and simple rules.

Membership in the fuzzy set of recalled rules decreases as the time since the last usage of the rule increases. Membership in the fuzzy set of applicable rules increases with the presence of supporting features and absence of detracting features. Membership in the fuzzy set of useful rules is defined by a weighting of its usefulness when last employed and long-term average usefulness. Membership in the fuzzy set of simple rules increases as the number of problem elements that must be considered to apply the rule decreases.

The model was used to predict the diagnosis behavior of 34 airframe and power plant mechanics when troubleshooting simulated automobile engines, turboprop engines, jet engines, turbocharged engines, autopilots, and helicopters. The model very closely predicted their overall troubleshooting performance. It predicted their exact choices of actions 50–60% of the time. The predicted actions were equivalent to humans' actions, in terms of membership in the choosable set, a much higher percentage of the time.

It should be noted that this type of model is inherently very context dependent. This is especially the case for the S-Rules. Hence, there is no general off-the-shelf model of human problem solving. There are generally applicable frameworks, but they must be populated with much context-specific knowledge. Consequently, there is no quick and easy way to computationally model the problem-solving behaviors of humans across domains.

Finance Theory

Finance theory as it relates to financial economics is concerned with relationships among financial variables, such as prices, interest rates, and shares. The two main areas of focus are asset pricing and corporate finance. The former is the concern of providers of capital, and the latter is the focus of users of capital. Finance theory addresses decision making under uncertainty, and closely relates to microeconomics and decision theory.

The field emerged in the late 1800s with innovators such as Wilfredo Pareto. Waiting, in effect, for the availability of computational power, the field began more rapid progress in the 1950s and 1960s with Kenneth Arrow and Georges Debreu's probabilistic model of markets, as well as Harry Markowitz's utility-based investment optimization. Another wave of innovation was associated with Fischer Black, Robert Merton, and Myron Scholes' option-pricing models. Now, of course, the enormous amount of easily available data is fueling many new ideas.

Two equations play a major role in the problems addressed in this book. The first is the standard production economics equation introduced in Chapter 6

$$\pi(t) = [P(t)Q_D(t) - VC(t)Q_P(t)] - FC(t) = \text{profit} \qquad (9.16)$$

P is price, Q_D and Q_P are quantities demanded and produced, respectively, VC is variable cost, and FC is fixed cost. The other standard and ubiquitous equations relate to discounted cash flow. Given a stream of profits π_i over a period *T*, the net present value of this free cash flow is given by

$$\text{NPV} = \sum_i \pi_i / (1 + \text{DR})^i \qquad (9.17)$$

where DR is the discount rate reflecting the time value of money, that is, amounts of money from future periods are discounted to the current period to reflect the earnings on the future amounts foregone while waiting, or the cost of borrowing the same amounts now for use until the future funds are available.

Typical assumptions associated with models based on financial theory include known projections of investments, operating costs, sales or revenues, other costs, and, by calculation, profit. These projections are often point estimates for each period of time, although most would readily agree that probability distributions at each point in time would be more meaningful. The aforementioned point projections enable calculating NPV.

If projections are seen as probability distributions and investment decisions are staged, with the decision to proceed at each stage being contingent on results to that point, we move into the realm of financial options. Then, we can calculate Net Option Value and Net Capital at Risk. The *Technology Investment Advisor* (Rouse et al., 2000; Rouse, 2007), discussed in Chapter 8, is a tool for projecting such metrics.

Figure 9.5 shows the models underlying the *Technology Investment Advisor*. The four models include an option-pricing model, *S*-curve market penetration model, production learning model, and competitive scenarios model. The interactions of these models vary depending on what aspects of the overall set of models are being employed. One can start at either end of this approach. One can use the *S*-curve and production learning models to project the financial information, which feeds the option-pricing model. Or, one can start with the option-pricing model to determine the option value of a given cash flow, and then use the *S*-curve and production-learning models to generate sales and cash flows, as well as "backcast" required R&D budgets to achieve these outcomes.

Note that Figure 9.5 indicates the Net Option Value as the central output of this set of models. In the absence of a contingent downstream investment decision, the analysis reduces to more traditional discounted cash flow methods. In the case, the output would be the usual Net Present Value. When applying this model-based approach to a portfolio of investments, it is quite likely that some investments will represent options and others will not. The *Technology Investment Advisor* was designed to handle such differences.

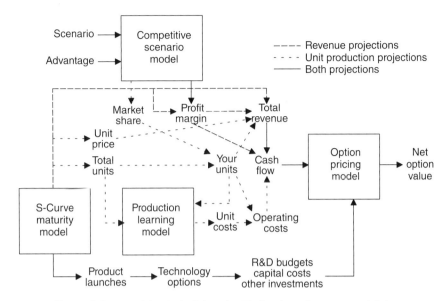

Figure 9.5 Models Underlying the *Technology Investment Advisor*

The inputs to the *S*-curve model include parameters for saturation, inflection, and time scale. For multiple product launches, the *S*-curve model also requires criteria for launching product derivatives, including parameters that characterize "descendent" *S*-curves. The *S*-Curve model projects either total market revenue or total market units sold each year, depending on whether production learning is being used. The production learning model uses the number of units sold each year and the initial unit cost to project the unit costs for each year.

Multiplying the unit cost for each year by the number of units yields the operating costs for each year. Subtracting the operating costs from the total revenue yields the free cash flow estimates needed by the option-pricing model. The number of product launches is used to determine the number of technology options required and subsequently the required R&D budgets. These budgets are coupled with capital costs and other investments to yield the total investment, or exercise price, needed by the options-pricing model. The combination of free cash flow and total investment enables determining the Net Option Value.

Note that Figure 9.5 indicates several paths for data flow and calculations. For instance, *S*-curve models can be used to project revenues or units. As another example, production learning can be driven by production volume or by the total market production volume – the former applies when production technology is proprietary, while the latter holds when all manufacturers buy the same production equipment from common vendors. For any particular analysis, one must decide which data flows and calculations are most appropriate.

Figure 9.5 also indicates the impacts of the competitive scenario model. Choices of competitive scenario, strategy advantage, and technology advantage result in projected upper and lower bounds for market shares and profit margins. One can split this difference or project other values of share or margin, depending on the particular competitive situation. It is easy to imagine that more sophisticated competition models, for example, based on game theory, may be feasible and warranted for some applications. However, such models would, by no means, be as "off the shelf" as the other models discussed in this chapter.

It is useful to consider how the integrated set of models depicted in Figure 9.5 address uncertainties, risks, and information precision. Random variations of projected cash flows are inherent to option-pricing models. The probabilistic nature of R&D – that is, whether viable options are created and whether they are adopted – is represented by technical and market success rates, respectively. The risks of competition for market share and margins – as well as other forces that could undermine share and margin – are represented via competitive scenarios. Of course, all of the parameters associated with the models in Figure 9.5 are only estimates and are subject to imprecision. The impact of this imprecision was assessed via sensitivity analysis and Monte Carlo analysis.

Summary

In this section, we discussed eight different theories and the paradigms they provide for modeling phenomena of interest. We discussed typical assumptions underlying each paradigm and the response variables usually predicted with each type of model. Common equations associated with each paradigm were presented and example applications illustrated.

I hasten to note that the eight paradigms presented here by no means exhaust the repertoire of models available. They do, however, represent commonly used approaches and courses on these topics are well represented in university curricula in engineering, computer science, management, and behavioral and social sciences.

LEVELS OF MODELING

Having reviewed a range of theories that we can draw upon, we are now ready to address Step 5: Identify Alternative Representations of These

Phenomena. We will pursue this is the context of the architecture of public–private enterprises introduced in Chapter 8 (Figure 8.8). In this section, we use this architecture to organize thinking about how to use the theories summarized in the last section for each of the layers in the enterprise architecture (Basole, et al., 2011) – see Table 9.3.

Note that the multilevel representation in Figure 8.8 differs significantly from the hierarchy of phenomena discussed in Chapter 3 in Figure 3.1. While there are some similarities between these two representations of levels of abstraction, they serve rather different purposes. Figure 8.8 is enterprise oriented and, therefore, lower levels are explicitly subsets of higher levels. In contrast, Figure 3.1 is intended to provide an organizing scheme for Chapters 4–7 and lower level phenomena are not inherently subsets of higher level phenomena in any formal sense.

The ecosystem models draw heavily upon Chapters 6 and 7, particularly macroeconomics and social system models. These models are often highly aggregated and used to predict overall economic metrics such as gross domestic product (GDP) growth and inflation. We seldom need, for instance, to model each economic actor's decisions to predict GDP growth and inflation. Nevertheless, these variables can be very important. For example, higher levels of inflation mean that downstream healthcare savings due to upstream prevention and wellness will be more valuable, although this does depend on the discount rate employed.

TABLE 9.3   Levels of Modeling, Example Issues, and Potential Models

Level	Issues	Models
Ecosystem	GDP, supply/demand, policy	Macroeconomic models
	Economic cycles	Dynamic system models
	Intra-firm relations, competition	Network models
Organizations	Profit maximization	Microeconomic models
	Competition	Game theory
	Investment	DCF, options
Processes	People, material flow	Discrete-event models
	Process efficiency	Learning models
	Workflow	Network models
People	Consumer behavior	Agent-based models
	Risk aversion	Utility models
	Perception progression	Markov, Bayes models

At the organization level, models from Chapters 5–7 are central, particularly microeconomic models and constructs from decision theory and finance theory. At this level, we are concerned with maximizing profits, making investment decisions, and dealing with competitors. Of course, all of these issues and decisions occur within the context of the "rules of the game" determined at the ecosystem level. The returns on investments in the process level below (in Figure 8.8) are also influenced by the potential payoffs from process improvements and new capacities.

Models employed at the process level draw heavily on Chapters 4–6. This level often involves substantial interactions of physical and human phenomena. This is where many work activities and tasks are accomplished to deliver capabilities and associated information. Simulation of these processes as well as their evolution over time, in part due to production learning, is often how the theories associated with these phenomena are instantiated and studied.

The people level draws upon the phenomena discussed in Chapters 5 and 7, as well as elements of Chapter 6. Here we are concerned with humans' perceptions, decisions, and actions, as well as how human health and possible disease states evolve. At this level, humans are both the agents of action and the objects of action through natural physical processes. It is also important to understand that the people level is both enabled and constrained by the capabilities and information provided by the process level.

REPRESENTATION TO COMPUTATION

Step 5 also involves translating the model-determined representations of phenomena into computational form. There are several common computational frames.

Dynamic Systems

The dynamic systems frame is concerned with transient and steady-state responses as well as stability of dynamic systems. Central constructs are stocks, flows, feedback, error, and control. Such systems are represented using differential or difference equations. Both have continuous states, but the former involves continuous transitions while the latter involves discrete transitions. Various elements of these models may be stochastic in nature, for example, disturbances and measurement noise. A range of tools can be used to solve such equations (see below), but the essence of the computation with all approaches is the relationship between future system states and past system states over time.

Discrete-Event Systems

The discrete event frame is concerned with steady-state responses in terms of average waiting time and time in the system, as well as average number of entities waiting and number of entities in the system. Central constructs are flows, capacities, and queues, as well as allocations of resources and time-based metrics. Such systems are represented using Markov chains with discrete states and continuous transitions. Also important are probability distributions associated with arrival flows (e.g., Poisson) and service processes (e.g., exponential). A variety of tools can be used to compute the responses of discrete-event systems (see below). As with dynamic systems, the essence of the computation with all approaches is the relationship between future system states and past system states, in this case averaged over time.

Agent-Based Systems

The agent-based frame focuses on large numbers of independent entities and the emergent responses of the collection of entities over time. Central constructs, for each agent, are information use, decision making, and adaptation over time. Such systems are represented by the sets of information sampling rules and decision-making rules assigned to each agent. These systems may also incorporate network models of relationships among agents. A range of tools can be used to compute the responses of agent-based systems (see below). The essence of all these approaches is simulation to compute the evolving state of each agent and collective patterns of behaviors. Of particular importance with such models is the notion that the observed emergent behaviors are not explicitly specified by the agents' rules.

Optimization-Based Frame

Another computational frame can overarch one or more of the above three frames. Beyond simply controlling a dynamic system, we may want to optimally control it as discussed earlier in this chapter. Beyond predicting the queuing time in a discrete-event system, we may want to optimally allocate resources to minimize some criterion that differentially weights the queuing times of different types of entities. Beyond observing the emergent behaviors of an agent-based system, we may want to design optimal incentive systems that maximize the likelihood of desirable emergent behaviors.

Thus, the problem to be solved using the models discussed in this chapter may be determination of the optimal control strategy, the optimal allocation of resources, or the optimal incentive system. Such aspirations are pervasive in operations research and systems engineering, as well as economics, finance, and so on. The Holy Grail is the "best" answer that maximizes benefits and minimizes costs.

We must keep in mind, however, that all of these pursuits must inherently solve their optimization problems in the contexts of models of the systems of interest rather the complex reality of the actual systems. Indeed, the achievement of the "best" answer can only be proven within the mathematical representations of the system of interest. To the extent that the mathematical models of the system are good approximations, the best answer may turn out to be "pretty good."

When there are humans in the system, we have to deal with not only our model of the system but also with humans' models of the system. Recall the notion of "constrained optimality" from Chapter 5. Succinctly, it is assumed that people will do their best to achieve task objectives within their constraints such as limited visual acuity, reaction time delays, and neuromotor lags. Thus, predicted behaviors and performance are calculated as the optimal behavior and performance subject to the constraints limiting the humans involved. As noted in Chapter 5, if these predictions do not compare favorably with subsequent empirical measurements of behaviors and performance, one or more constraints have been missed (Rouse, 1980, 2007).

Determining the optimal solution for any particular task or tasks requires assumptions beyond the likely constraints on human behavior and performance. Many tasks require understanding of the objectives or desired outcomes and inputs to accomplish these outcomes, as well as any intervening processes. For example, drivers need to understand the dynamics of their vehicles. Pilots need to understand the dynamics of their aircraft. Process plant operators need to understand the dynamics of their processes. They also need to understand any trade-offs between, for example, accuracy of performance and energy required, human and otherwise, to achieve performance.

This understanding is often characterized as humans' "mental models" of their tasks and context. To calculate the optimal control of a system, or the optimal detection of failures, or the optimal diagnoses of failures, assumptions are needed regarding humans' mental models. If we assume well-trained humans who agree with and understand task objectives, we can usually argue that their mental models are accurate, for example, reflect the actual physical dynamics of the vehicle.

They key point is that optimizing the types of systems addressed by the archetypal problems in this book requires that we develop models of humans' model of reality. This is tractable if there are enough constraints on humans' choices and behaviors. Without sufficient constraints, however, the whole notion of optimality can become extremely ambiguous and often useless.

Summary

Most of the applications of the aforementioned computational frames have involved modeling and representation of the "physics" of the environment, infrastructure, vehicles, and so on. These are certainly important elements of many overall multilevel models. However, the greatest challenges in developing such models for the types of problems addressed on this book are modeling and representation of the behavioral and social activities and performance throughout the system, especially when it cannot be assumed that the human elements of the systems will behave in accordance with the objectives and "rules of engagement" of the overall system.

Chapter 4–7 provides a wealth of possible representations of behavioral and social phenomena, but these approaches are nonetheless subject to much more human variability than experienced for physical systems, especially systems that were designed or engineered. This is not a cause for despair, but a caution that human variability may be as important, or more important, than average human responses. This variability can, for example, result in deteriorating overall systems performance despite the fact humans, on average, are performing just as expected.

MODEL COMPOSITION

Step 6 of the methodology – Assess the Ability to Connect Alternative Representations – involves assessing the ability to connect alternative representations. This is also termed model composition. There are two overarching model composition issues: dealing with entangled states and assuring consistency of assumptions (Pennock & Rouse, 2014a, 2014b).

Entangled States

Model states are entangled when the state of one model depends on the state of another model and vice versa. This may require that the two models be solved simultaneously. However, this requirement may unacceptably increase the level of computational complexity.

Consider a multitask situation involving continuous control and randomly arriving discrete tasks. Flight management is a good example of such a task. The pilot has two tasks: guiding the aircraft along a displayed flight path and responding to discrete events such as communications with air traffic control and investigating sources of alarms.

The aircraft guidance task is well modeled using an optimal preview control model, which is an extension of the optimal control model discussed earlier in this chapter. Responding to discrete events can be modeled as a simple queuing problem. The question is how to compose these two models.

Answering this question involved a principled heuristic that did not violate the assumptions of either model. The idea was to schedule the discrete tasks when the computed optimal control was slack. We then recomputed the optimal preview control with infinite (very large) weighting on energy used during slack periods. This resulted in the optimal redistribution of control effort to handle the slack periods, which compared quite favorably with actual pilot behaviors.

Consider the archetypal problem of congestion pricing. This involves a partial differential equation model of traffic flow and congestion, composed with agent-based models of drivers' decision making based on perceptions of traffic flow and congestion. The causal chain is projected traffic flow to congestion pricing to drivers' reactions to actual traffic flow. In the real world, this dynamic would probably lead to an eventual equilibrium with prices adjusted to reflect drivers' choices.

However, this dynamic could also lead to a bubble as it did for the archetypal financial system problem. The entangled states between financial instruments and real estate prices entered a positive feedback loop that pushed up housing prices until the bubble burst. In contrast, we do not see traffic bubbles bursting because the system is much more observable and controllable. Traffic engineers can adjust pricing rather than having to wait for the market to do this.

For the flight management problem, we found a way to disentangle the states of the two models. For the congestion-pricing problem, we expect that the interactions between pricing decisions and drivers' decisions would eventually reach equilibrium. For the financial system problem, we would expect an emerging positive feedback loop to lead to a bubble that would eventually burst.

What we want to avoid is a modeling composition bubble. We do not want to predict divergent behavior that could not happen in reality. One suggestion is to approach this issue empirically. Leave the states entangled, letting the two models react to each other. See whether equilibrium or a bubble emerges. Determine which model variables most affect these phenomena. Determine whether these possibilities reflect physical realties or are just modeling artifacts. If they are artifacts, look for a principled heuristic to disentangle the models.

Consistency of Assumptions

Step 7 of the methodology – Determine a Consistent Set of Assumptions – concerns determining a consistent set of assumptions for the set of representations being computationally composed. Basic consistency issues include time scales, units of measure, and coordinate systems. It is, of course, important to address such incompatibilities. There are, however, several other consistency issues that can be rather subtle.

Input Inconsistency   Most queuing models assume independent arrivals of entities for service, yet many processes have scheduled arrivals, for example, healthcare delivery and air traffic control. The former models will overestimate variability, while the latter will underestimate variability. The empirical truth is usually somewhere is the middle.

Output Inconsistency   Consider a composition where one model projects a transient response while another projects steady state. For example, one model predicts wind velocity by the hour while the other projects average temperature for the day. Another example is one model predicting the price of a share of stock by the hour and another predicting the average closing index for the day. If one were aware of such incompatibilities, one would work it out. But, what are the implications of not knowing that the composition has this type of problem?

Process Inconsistency   Similar Markov assumptions underlie typical dynamic systems, control theory, estimation theory, and queuing theory models, that is, the future state only depends on the current state. What happens when we compose one of these models with, say, an econometric autoregressive time series model that assumes past states, as well as the current state, influences future states? If the econometric model is correct (i.e., past states matter), then the Markov assumption is wrong. What are the implications?

Parameter Inconsistency   Consider another composition where one model assumes no friction to simplify calculations while another depends on friction to dissipate vibrations. An engine might be a good example. The fuel injection system model assumes no friction, while the crankshaft-pistons model assumes friction to dampen vibrations at high RPM. This might be acceptable if the two functions are, in effect, physically separable.

Separability Principle   In some cases, two models can be solved independent of each other. A good example is the Kalman filter and optimal controller for the LQG problem discussed earlier in this chapter. Of course, the assumptions underlying the two models in this case are fully consistent. Separability is an issue of importance in several areas of science and mathematics.

Observations

Entangled states require that we address them in some manner. The control plus queuing example represents a principled heuristic – it does not violate either model's premises. The congestion example might not be a problem at all if the congestion pricing and driver behaviors reach equilibrium. We need to approach this in a way that avoids model-induced phenomena that are unlikely to happen in the real world.

Inconsistent assumptions, if one is aware of them, may or may not pose problems. They may be relatively harmless, especially for cases where separability can be argued. However, making assumptions consistent may be intractable. Then, the question is how much the inconsistency costs in terms of erroneous predictions and misleading insights. There is no general context-free answer to this question.

COMPUTATIONAL TOOLS

This is a wide range of commercially available software tools for creating computational instantiations of the theories and models discussed in this chapter. Many of these packages have been available for quite some time, have been updated regularly, and have sizable user groups and online forums for support.

Mathematica and MATLAB are both commercial software packages for supporting representation and solutions of a wide range of equations. This, of course, requires that one explicitly derive the equations to be solved. There is a range of other software packages that support one or more of the theories or paradigms discussed earlier in this chapter and, in effect, derive the needed equations from a graphic portrayal of the model.

Simcad Pro and VisSim are both commercial software packages for supporting continuous simulation. They have block diagram–oriented graphical user interfaces. They both provide options for numerical integration techniques, as they are solving continuous differential equations rather than discrete difference equations.

Arena and Simio are both commercial software packages for supporting discrete-event simulation of network models. These models typically involve queuing of entities that arrive at entry nodes, wait for service, and are then serviced and passed on to the next node until the whole servicing process is complete. These packages include capabilities for analysis of simulation runs and graphic portrayal of results.

NetLogo and Repast both support agent-based modeling and simulation for large number of agents. Model developers provide behavioral rules to each agent, and the simulation enables seeing the overall system behaviors that emerge from the collective actions of all agents. Both tools are free.

Stella and Vensim are both commercial software packages for supporting systems dynamics modeling using Jay Forrester's framework (Sterman, 2002). These tools have easy-to-use graphical interfaces and a variety of associated analytical capabilities.

AnyLogic is a commercial software package that simultaneously supports discrete-event simulation, agent-based simulation, and systems dynamics modeling. The ability to pass data among these three representations enables multilevel modeling of complex phenomena. Its easy-to-use graphical interface enables rapid prototyping of initial models.

R is a statistical package that also supports simulation but does not provide functionality tailored to discrete-event simulation, agent-based simulation, or systems dynamics modeling. Nevertheless, one can embed equations from these paradigms in an R simulation. R is free.

ModelCenter is a commercial software package for integrating outputs for independent software models. It does not provide functionality tailored to discrete-event simulation, agent-based simulation, or systems dynamics modeling. It also includes capabilities to support team projects.

It is useful to note that some of these tools are often more useful for prototyping versus creating the final versions of model-based tools as exemplified by *Product Planning Advisor* and *Technology Investment Advisor*. There are two reasons. First, it may be that the user interface possibilities supported by the package do not match the needs associated with, for example, a policy flight simulator in the *Immersion Lab*. Second, some of the packages are unacceptably slow for large problems, in part because they have so much infrastructure support devoted to being easy to use. This has sometimes resulted in our proving the idea using the commercial package and then recoding in a lower-level language (e.g., C++, Java, Python) to enable the simulation to run much more quickly.

CONCLUSIONS

This chapter addressed Steps 5–7 of the overall methodology. We first summarized the phenomena associated with the six archetypal problems discussed throughout this book. Modeling paradigms potentially useful for addressing these phenomena were then discussed, including dynamic systems theory, control theory, estimation theory, queuing theory, network theory, decision theory, problem solving theory, and finance theory.

A multilevel modeling framework was used to illustrate how the different modeling paradigms can be employed to represent different levels of abstraction and aggregation of an overall problem. The next consideration was moving from representation to computation. This included discussion of model composition and issues of entangled states and consistency of assumptions. This chapter also provided a brief overview of software tools available to support use of these computational approaches.

We now move on to consider a wide range of case studies using the material presented in Chapters 1–9. These case studies are cast as problem-solving endeavors in that the overarching goals were seldom, if ever, to develop computational models of complex domains. The motivations included new product planning in highly competitive markets, technology portfolio investments in very uncertain markets, and enterprise transformation to counter threats to existing business models.

REFERENCES

Arrow, K.J. (1951). *Social Choice and Individual Values*. New York: Wiley.

Athans, M. (Ed.). (1971). Special issue on the linear–quadratic–Gaussian problem in control system design. *IEEE Transactions on Automatic Control*, AC-16 (6), 529–552.

Basole, R.C., Rouse, W.B., McGinnis, L.F., Bodner, D.A., & Kessler, W.C. (2011). Models of complex enterprise networks. *Journal of Enterprise Transformation*, 1 (3), 208–230.

Batty, M. (2013). *The New Science of Cities*. Cambridge, MA: MIT Press.

Bellman, R.E. (1967). *Introduction to the Mathematical Theory of Control Processes*. New York: Academic Press.

Erlang, A.K. (1909). The theory of probabilities and telephone conversations. *Nyt Tidsskrift for Matematik*, 20 (B), 33–39.

Kalman, R. (1960). A new approach to linear filtering and prediction problems. *Transactions of the ASME, Journal of Basic Engineering*, 82, 34–45.

Kalman, R.E., (1961). On the general theory of control systems, *Proceedings of the 1st International Congress of IFAC*, Moscow, p. 481.

Keeney, R.L., & Raiffa, H. (1976). *Decisions With Multiple Objectives: Preference and Value Tradeoffs*. New York: Wiley.

Morse, P.M. (1958). *Queues, Inventories and Maintenance*. New York: Wiley.

Nash, J.F. (1950). Equilibrium points in *N*-person games. *Proceedings of the National Academy of Sciences of the United States of America*, 36, 48–49.

Newell, A., & Simon, H.A. (1972). *Human Problem Solving*. Englewood Cliffs, NJ: Prentice Hall.

Pennock, M.J., & Rouse, W.B. (2014a). The challenges of modeling enterprise systems. *Proceedings of the 4th International Symposium on Engineering Systems*. Hoboken, NJ, June 8–11.

Pennock, M.J., & Rouse, W.B. (2014b). Why connecting theories together may not work: How to address complex paradigm-spanning questions. *Proceedings of the 2014 IEEE International Conference on Systems, Man and Cybernetics*, San Diego, October 5–8.

Pontryagin, L.S. (1962). *The Mathematical Theory of Optimal Processes*. New York: Wiley.

Rasmussen, J. (1986). Information Processing and Human-Machine Interaction. New York: North-Holland.

Rasmussen, J., & Rouse, W.B. (Eds.). (1981). *Human Detection and Diagnosis of System Failures*. New York: Plenum Press.

Rasmussen, J., Pejtersen, A.M., & Goodstein, L.P. (1994). *Cognitive Systems Engineering*. New York: Wiley.

Rouse, W.B. (1980). *Systems Engineering Models of Human-Machine Interaction*. New York: North Holland.

Rouse, W.B. (1983). Models of human problem solving: Detection, diagnosis, and compensation for system failures. *Automatica*, 19(6), 613–625.

Rouse, W.B. (1991). *Design for Success: Human-Centered Design of Products and Systems*. New York: Wiley.

Rouse, W.B. (2007). *People and Organizations: Explorations of Human-Centered Design*. New York: Wiley.

Rouse, W.B., & Howard, C.W. (1995). Supporting market-driven change, in D. Burnstein, Ed., *The Digital MBA* (pp. 159–184). New York: Osborne McGraw-Hill.

Rouse, W.B., Howard, C.W., Carns, W.E., & Prendergast, E.J. (2000). Technology investment advisor: An options-based approach to technology strategy. *Information Knowledge Systems Management*, 2 (1), 63–81.

Rouse, W.B., & Serban, N. (2014). Understanding and managing the complexity of healthcare. Cambridge, MA: MIT Press.

Sheridan, T.B. (1992). *Telerobotics, Automation and Human Supervisory Control*. Cambridge, MA: MIT Press.

Sheridan, T.B., & Ferrell, W.R. (1974). *Man-Machine Systems: Information, Control, and Decision Models of Human Performance*. Cambridge, MA: MIT Press.

Simon, H.A. (1978). How to decide what to do. *The Bell Journal of Economics*, 9 (2), 494–507.

Sterman, J.D. (2002). Systems dynamics modeling: Tools for learning in a complex world. *California Management Review*, 43 (4), 8–25.

Von Neumann, J., & Morgenstern, O. (1944). *Theory of Games and Economic Behavior*. Princeton, NJ: Princeton University Press.

Zadeh, L.A. (1965) Fuzzy sets. *Information and Control*, 8 (3) 338–353.

10

PERSPECTIVES ON PROBLEM SOLVING

INTRODUCTION

We use models to answer questions and solve problems. Often these problems involve designing solutions in terms of physical form, functional capabilities, and policies intended to incentivize or inhibit particular behaviors. The key point is that models are intended to support problem solving.

Initially, our models are limited to visualizations, perhaps just sketches. These visualizations may evolve to become more elaborate and perhaps interactive. They enable exploration of connections among entities and how relationships among entities work. Such visualizations are models. They express the problem solver's perceptions of what phenomena matter, how they interact, and key trade-offs.

Sometimes, a good visualization is all that is needed. The problem-solving group's discussion and exploration of the visualization lead to a conclusion on how to proceed. In other situations, deeper explorations are needed. These explorations may involve more formal representations of phenomena and relationships. Deep computation may be warranted, but perhaps used sparingly.

The discussions and explorations usually lead to creative suggestions for possible courses of action. All the creativity comes from the group of problem

Modeling and Visualization of Complex Systems and Enterprises:
Explorations of Physical, Human, Economic, and Social Phenomena, First Edition. William B. Rouse.
© 2015 John Wiley & Sons, Inc. Published 2015 by John Wiley & Sons, Inc.

solvers, not computers. In other words, policy flight simulators seldom fly themselves. Instead, computers provide the means to explore the implications of seemingly good ideas. Bad ideas are rejected and good ideas are refined, perhaps for empirical evaluation.

This whole process results in many graphs and perhaps surface plots. Models may be revised and extended. The most powerful impact, however, is that the problem-solving group develops a shared mental model of what effects what, what really matters, and what trade-offs are crucial. Members of the large number of groups with whom I have worked (elaborated below) have repeatedly told me that the resulting shared mental model was far more powerful than any of the graphs or plots produced.

This chapter brings all the material in this book together to discuss a large set of case studies involving well over one hundred enterprises and several thousand participants. Discussion of these case studies focuses on how problem solving was addressed, the roles that interactive models played in problem solving, and the types of insights and decisions that resulted. This chapter concludes with consideration of key research issues that need to be addressed to advance the approach to problem solving advocated in this book.

WHAT IS? VERSUS WHAT IF?

It is useful to begin by distinguishing two very different types of overarching questions. Understanding the nature of problems often involves assessing "what is." This is the realm of "big data." In contrast, solving problems often also involves asking "what if." This typically involves exploring possibilities that have never been tried before. Visualization and computational models are used to predict the consequences of such possibilities.

How can such predictions be validated? A limited type of validation can be conducted by predicting the performance of the status quo system. If the models fail this test, something is wrong somewhere. However, full empirical validation is inherently not possible. Thus, the predictive validity of the models will remain in question. How can we really know that our predictions will come to be true?

The simple answer is, "We cannot." Nevertheless, this is not a fatal flaw if the goal is insights rather than predictions. If the goal is to understand why ideas might work or not – might be good ideas or bad ideas – then the models can be used to provide prediction-based insights. We can learn what *might* happen and the conditions under which these possibilities are likely to happen. We might then try to influence these conditions.

Experience has shown that this is a good way to quickly discard bad ideas and to refine good ideas for subsequent empirical evaluation. We can determine why seemingly good ideas are actually bad. We can also learn the conditions under which apparently good ideas are likely to succeed. Such insights are the hallmarks of creative and successful problem solving.

CASE STUDIES

In Chapter 8, we discussed the results of 100 engagements that included over 2000 executives, senior managers, and disciplinary specialists involved in strategic planning and new product planning (Rouse, 1998). Much more recently, the construct of a policy flight simulator has been elaborated and illustrated (Rouse, 2014). The findings of both of these endeavors were discussed in Chapter 8.

In this section, we will move beyond issues of visualization and modeling to discuss the nature of problem solving encountered in a large number of engagements involving almost 40 enterprises. Most of these enterprises were Fortune 500 companies, but a few were government agencies and healthcare providers. All of these enterprises made real and substantial decisions using the tools and simulators discussed in this book.

The objectives of these engagements included the following:

- Business planning to address market and technology opportunities
- New product planning in highly competitive markets
- Technology portfolio investments in very uncertain markets
- Enterprise transformation to counter threats to existing business models

Business Planning

Our primary offerings focused on new product planning and technology investment analysis. However, many customers asked us to help them with developing strategic plans for their overall enterprise. We expanded our portfolio of tools to include **Business Planning Advisor** and later **Situation Assessment Advisor** to support these engagements. Other than embedded spreadsheet capabilities, these tools relied on embedded "expert systems" to assess the strengths and weaknesses of plans, as well as the situational premises upon which the plans were based (Rouse, 1992, 1996; Rouse & Howard, 1995). This section presents 4 cases that addressed particular strategic issues pursued by 12 enterprises.

Products versus Services Two companies – in one case an airplane company and the other a satellite company – were concerned that their products earned lower profit margins than those of providers of services to owners of their products. In fact, the service providers enjoyed both much higher profit margins and much lower invested capital. An executive in one firm said, "Why don't we get someone else to produce the airplanes?" After many planning discussions and debates, the airplane company started a services subsidiary and the satellite company spun off the services business. The financial projections were attractive, but both companies felt that the overhead costs of their enormous infrastructures would overly burden the costs of service offerings. They also wondered whether their companies' cultures could really become service oriented.

Defense Conversion   Two companies – one in electronics and the other in munitions - were seeing their defense revenues declining. They developed business plans for other market sectors, for example, healthcare technology. Their detailed plans showed strong possibilities if they could escape their defense technology mentalities. Interestingly, the munitions company concluded that their customers did not really want "bombs and bullets," but had no other way to exert force. This insight led to many creative ideas from R&D, which were rejected by senior management as not realistic.

R&D as a Service   Four organizations – a pharmaceutical company, an aerospace company, and two government labs, one in the United States and the other international   felt that they needed to be more customer-oriented in their R&D offerings, both internally and externally. They wanted to focus on value to customers. This seemed reasonable until they all discovered that their cultures were oriented toward gaining and maintaining budgets (i.e., inputs) rather than contributing to solving customers' problems (i.e., outputs or outcomes). This led to a shift from thinking in terms of their technology portfolios to articulating their offerings in terms of the types of problems they could solve for customers. In other words, they moved from talking about "cool stuff on the shelf" to contributions to competitiveness of their customers' offerings, costs of their customers' solutions, and so on.

Academic Strategies   Four universities – two private and two public – wanted to think more strategically about their offerings and value propositions. This required thinking about future scenarios and their competitive positions in these scenarios. Four alternative future scenarios were developed, one of which was "business as usual" while the others were very different in terms of students served, methods of delivery, and roles of faculty members. Everyone intellectually embraced the idea of the four scenarios, but rank and file faculty members had great difficulty discussing anything other than business as usual. Many great discussions ensued and coherent well-articulated strategic plans were created. An interesting aspect of all four plans was the underlying, unspoken intent of making sure that all constituencies embraced the plan. This tended to result in lofty, but plain vanilla plans. These plans did not result in curtailing any existing operations; hence resources were scarce for undertaking any new initiatives.

Summary These four examples serve to illustrate several types of issues in strategic planning. The first two examples showed how legacy commitments often constrain future plans. Many executives have told me that their toughest problem is running the company they have while they try to create the company they want. The third example depicts the difficulties of becoming truly customer oriented. Large, well-established enterprises often evolve to the point that they feel it is customers' responsibility to keep them profitable in business. The final example demonstrated the difficulties of thinking strategically amidst a wide range of constituencies with varied, and sometimes conflicting, interests. These four examples illustrate the difficulties of both thinking *and* acting strategically. Not every enterprise encountered these difficulties but they are quite common.

New Product Planning

We conducted a very large number of new product planning engagements with a wide range of enterprises. In this section, I highlight four of these experiences. Before discussing these cases, it is interesting to note a few observations from other cases.

Working with an aircraft engine company, we discovered that engineering and marketing disagreed about the units of measure of customers' key variables. This emerged because the *Product Planning Advisor* (PPA) asks users to provide units of measure for all attributes. This was surprising because the leaders of these two functions had worked together for many years. They commented that they never had any reason in the past to discuss units of measure.

A semiconductor company was listed in the *Guinness Book of Records* for the speed of their microprocessors. Every product planning engagement with them included the objective that they retain their place in the record book. This was held of highest importance even for applications where the increased speed of the microprocessor provided no benefits to customers, due to limitations of the other elements of the system. They later relented on this objective, as one of the cases in the following text discusses.

We worked with a chemical company planning new pesticide and herbicides. They were required to test these products on rats to assure that they were not carcinogenic. They reported that none of the rats had gotten cancer because all of them died immediately upon ingesting the chemicals. However, this was inconsistent with the required testing protocol. This is one of the most unusual product planning engagements I ever experienced.

A British aerospace company acquired our product planning methods and tools. They were concerned about product support. We guaranteed them that any problem encountered or question that emerged would be solved or answered within 24 h. We met this commitment and the Managing Director, at a public industry meeting, commented that this quality of service was amazing and that he never before had experienced such responsiveness. The Internet and email enabled all of this.

The following four examples of how the PPA has been used illustrate the ways in which this tool is applied and the types of insights that are gained. In particular, these examples depict trade-offs across stakeholders and how the impacts of assumptions can be explored. It is important to note that these examples show how product-planning teams have reached counterintuitive conclusions using PPA. However, use of PPA does not, by any means, always result in such conclusions.

Automobile Engine   A team working on new emission control systems decided to evaluate an earlier technology investment using PPA. They compared the chosen approach to four other candidates that had been rejected with the earlier decision. Development and use of the market/product models resulted in the conclusion that the chosen approach was the worst among the five original candidates. This surprising conclusion led to in-depth exploration of the assumptions built into their PPA models. This exploration resulted in further support for these assumptions. Reviewing these results, the team leader realized that the earlier decision had not fully considered the impact of the alternatives on the manufacturing stakeholder. The earlier choice had been of high utility to customers and other stakeholders, but was very complex to manufacture. As a result of this insight, a new approach was adopted.

Microprocessors   A major competitor in the semiconductor market was planning a new high-end microprocessor. They were very concerned with time to market, worried that their next generation might be late relative to the competition. Their planning team included people from engineering, manufacturing, marketing, and finance. Using PPA, they found that time to market was critical, but it was not clear how it could be significantly decreased. One of the manufacturing participants suggested a design change that, upon analysis, would get them to market a year earlier. The team adopted this suggestion. He was asked, "Why have you never suggested this before?" He responded, "Because you have never invited manufacturing to these types of meetings before." Product planning with PPA often results in involvement of a richer set of internal stakeholders.

Digital Signal Processor   The product planning team began this effort convinced that they already knew the best function/feature set with which to delight the market. The marketing manager, however, insisted that they test their intuitions using PPA. After developing the market/product models and using them for competitive analyses, the team concluded that assumptions regarding stakeholders' preferences for three particular attributes, as well as the values of these attributes, were critical to their original intuitions being correct. Attempts to support these assumptions by talking with stakeholders, especially end users and customers, resulted in the conclusions that all three assumptions were unsupportable. The team subsequently pursued a different product concept.

Medical Imaging System   A product planning team had developed an advanced concept for medical imaging that they argued would enable their company to enter a very crowded market, where a couple of brand name companies currently dominated. They used PPA to assess the market advantages of their concept relative to the offerings of the market leaders. Initial results showed a considerably greater market utility for their potential offering. Attention then shifted to the likely reactions of the market leaders to the introduction of this advanced product. The team's expectation was that the leaders would have to invest in 2 years of R&D to catch up with the new technology embodied in their offering. However, using the "How to Improve?" feature for PPA models of the competitors' offerings resulted in the conclusion that the best strategy for the market leaders was to reduce prices significantly. The team had not anticipated this possibility – someone said, "That's not fair!" This caused the team to reconsider the firmness of their revenue projections, in terms of both number of units sold and price per unit.

Summary   These four examples serve to illustrate several types of issues in new product development. The first example showed how the concerns of a secondary stakeholder could affect the attractiveness of a solution. The second example illustrated how a planning team gained insights via the discussions and debates that this tool engenders. The third example depicted the impact of unsupportable assumptions regarding the preferences of primary stakeholders. The final example demonstrated how the likely reactions of competitors impact the possible market advantages of a product or system. Taken together, these four examples clearly illustrate how a human-centered orientation helps to avoid creating solutions that some stakeholders may want but other stakeholders will not support or buy.

Technology Investments

The *Technology Investment Advisor* (TIA) emerged from the confluence of two efforts, one for a government agency and another for a large electronics company. Both enterprises wanted to bring greater rigor to their R&D investment processes. The common conclusion was that the purpose of R&D is to create "technology options" for operational or business units to exercise if and when needed (Rouse & Boff, 2004). This conclusion led to a technology investment planning process and the TIA. We used this method and tool extensively with over 10 enterprises – two government agencies and the rest private sector technology-oriented companies. The 4 case studies discussed in this section are as subset of the over 30 engagements on this topic. Fourteen of these cases are discussed by Rouse and Boff (2004).

MRAM   Magnetoresistive random-access memory stores data by magnetic storage elements rather than by electric charge or current. Thus, it does not depend on the availability of electricity. We worked with the research arm of a large electronics company to project the impact of their investments in this technology on future revenues and profits. Using TIA, we projected several generations of this technology via a family of S-curves, along with decreasing unit costs due to production learning. The projected Net Option Value for the $20 million R&D budget requested was $450 million. This so impressed the CEO that he doubled the R&D budget, requesting a plan for how to spend the extra monies well. The launching of the resulting MRAM products led to a multibillion-dollar division of this company.

Optical Computing   This technology involves using photons rather than electrons for computing. The large electronics company (different from the MRAM company) with whom we worked already had two optical computing product lines that were addressing rather narrow markets. We argued that their existing technology portfolio gave them "technology options" for entering new, broader markets. This required stretching their planning horizon to allow for a sufficient number of S-curves to yield substantial production learning benefits, which resulted in impressive Net Option Values. Of most significance, the CEO of one subsidiary sent the CFO of the corporate parent (our customer) an email indicating that he was much more comfortable with the new financial projects but, much more importantly, they now had a transformed sense of the business they were in.

Unmanned Air Vehicles   A defense agency outside of the United States was debating the use of unmanned air vehicles versus microsatellites for a reconnaissance and surveillance mission. Both required roughly the same R&D investments. Could we use both TIA and PPA to recommend which investment to make? The planning horizon was 20 years, with *S*-curves quite stretched compared to the aforementioned examples and minimal production learning due to low volumes. Nevertheless, the Net Option Values were sufficient to justify making both R&D investments, especially since the risks of both projects were quite different.

A difficult aspect of this project was the need to meet mission requirements immediately, despite the proposed R&D efforts requiring several years. They decided to meet the immediate need with manned aircraft. However, this would result in having unneeded aircraft assets after several years. How could we justify the additional investment to replace this capability? This led to an initiative to "sell" the aircraft to another agency once the new technology could support the mission. Once we obtained commitment from a "buyer" and could project the aircraft being "off the books," the net benefits of the new technologies were substantial. Interestingly, the operating costs of the manned aircraft were very high compared to the proposed technologies. Consequently, the manned aircraft would be an inferior solution, even if the purchase costs of the aircraft were zero.

Licensing Technology   Three companies with whom we worked – two large electronics companies and a large beverage company – were interested in using TIA to project the option values of licensing technologies. The two electronics companies had invested in creating options that other companies wanted to purchase. This required that we project the option exercise price for the licensee rather than the originator of the option. This would enable determination of a reasonable option purchase price – higher exercise prices dictate smaller purchase prices to make investments attractive. This led these companies to the idea of helping the licensees exercise the options. This resulted in additional consulting revenue for the two companies while also justifying higher licensing revenue, in conjunction with substantially lowering the licensees' risks.

The large beverage company was entertaining the purchase of an exclusive license to a technology that would enable them to meet new environmental regulations outside the United States. Without this technology, their revenues would be significantly decreased by the anticipated regulations. The financial benefits of owning this option (license) were straightforward to project, but the customer added an unusual wrinkle. The exclusive license they were seeking would prohibit their competitors from meeting the regulations, with substantial negative financial impacts. Once this became apparent and appeared in the news, our customer intended to provide use of the technology to their competitors free of charge, and expected to earn great praise for saving the day for the industry. Attaching economic value to this was beyond our expertise, but the company was quite comfortable making such estimates in terms of the value of free marketing and media time.

Summary   These four examples serve to illustrate several types of issues in addressing technology investments. The first example showed how a long-term perspective, including use of extensive company data on *S*-Curves and production learning, can lead to deep understanding of potential substantial returns on investments. The second example illustrated how a planning team gained insights into alternative markets for their technologies, in this case much larger and more profitable markets than they were currently serving. The third example depicted some of the subtleties of replacing a legacy investment with new technologies especially when the incumbent technology is not obsolete but much too costly to operate. The final example demonstrated the change in perspective needed when one is selling rather than buying technology options. Taken together, these four examples clearly illustrate how options-based thinking can change the nature of business models, particular in the sense of attaching value to what *might* happen and how one can be prepared to exploit the opportunity *if* it happens.

Enterprise Transformation

Thus far, we have discussed a dozen case studies of business planning, new product planning, and technology investments. These cases were not focused on totally redefining the enterprise, although there were certainly some implications for this. This section focuses on four cases where the status quo is being challenged and the enterprise has to entertain substantial fundamental change. Understanding of these cases can be pursued in much greater depth in Rouse (2006, 2011), and Rouse and Serban (2014).

Value Opportunities   It is much easier to change in pursuit of market and technology opportunities rather than in reaction to crises (see below). This case involves three companies – one in consumer products, another in electronics, and a third in supply chain services – who decided to move far beyond their founding vision. They accomplished this via acquisitions, buying high-margin smaller companies, high-technology start-ups, and companies that were best in class for key processes that the larger aspirations required. These companies took a very disciplined approach to digesting acquisitions. One of them told me, "Acquisitions represent a business process, not just a transaction." Thus, they carefully mapped the business processes of companies being acquired and then thoughtfully designed how these processes would dovetail with those of the larger enterprise. Successful acquisitions can be elusive, but these three companies figured out how to design for success.

Value Deficiencies The transformation of these three companies – one in aerospace, one in clothing, and a third in electronics – was motivated by crises. The aerospace company had acquired three aircraft companies without rationalizing processes and leadership, resulting in them competing against each other in a declining market. The clothing company had homogenized product design and marketing to the extent that brand identities had been lost, market share had substantially declined, and the financial condition of the company was abysmal. The electronics company had clung to a fading business model, in part due to arrogance, while their competitors had passed them by and were gaining market share.

All three companies had to first escape the delusion that the world had not changed. They then had to map and rationalize processes. As they did this, they dramatically thinned their executive ranks, along with employees at all levels. They focused intensively on understanding their "as is" and "to be" enterprises in terms of customers, efficiency, culture, and execution. Accountability became a hallmark, with much more frequent reports and reviews. Over several years, they came back to strong revenues and profits, as well as high regard in the stock market.

Population Health Chronic disease accounts for an increasingly large portion of the US healthcare costs. Chronic disease management, as well as prevention and wellness, can control these costs via low-cost interventions that keep patients out of emergency rooms and hospital beds, while also keeping people healthier. We worked with two large healthcare providers to evaluate and redesign programs for prevention and wellness, as well as chronic disease management, focused on hypertension, diabetes, and heart disease. We worked with a third provider focused on Alzheimer's disease.

We created computational models and interactive visualizations to explore ways in which to stratify patients (e.g., high vs. low risk) and the design of processes tailored to each stratum. Models were also developed to optimize staffing, scheduling, and routing practices. We found that the model building process often ended up driving process design. More specifically, we began each effort with processes supposedly designed and operational. As we mapped processes and simulated them, the customers immediately identified weaknesses and started to redesign these delivery processes. Thus, we ended up with iterative model-design cycles, which led to substantial improvements in outcomes for patients and decreased costs of delivery.

Affordable Care Act   Successful delivery of population health results in declining hospital revenues due to decreasing emergency room visits and in-patient admissions. These free up hospital capacities that have to be either redeployed or liquidated. Most hospitals are addressing this trend by attempting to serve larger catchment areas. In other words, if a smaller percentage of patients are going to go to the hospital, then the hospital needs to serve a larger population. However, there are not large numbers of unserved patients, especially profitable patients.

Thus, one outcome of the Affordable Care Act has been an intense interest in mergers and acquisitions among healthcare providers. In fact, as of this writing, mergers and acquisitions in the healthcare sector are much greater than in any other sector of the economy. To address this phenomenon, we created a model-based interactive visualization of the 44 hospitals in New York City. Each hospital in this agent-based simulation has an Income Statement, Balance Sheet, operational performance metrics, and market shares of diagnostic markets addressed.

This simulation results in hospitals making and receiving acquisition bids as providers try to increase revenues and profits, as well as identify turnaround opportunities. For all scenarios, roughly 20 hospitals disappear over the next 10 years, not as healthcare facilities but as independent enterprises. Interestingly, the hospital enterprises remaining vary as who acquires who is very path dependent. This interactive environment is being used to attract senior management to dialogs about alternative futures.

Summary   These four examples serve to illustrate several types of issues in addressing enterprise transformation. The first example showed how three companies systematically pursued acquisitions and designed processes to enable successful pursuit of new, broader aspirations. The second example illustrated the consequences of waiting until crises precipitated needs to transform, with enormous resources wasted and jobs eliminated. The third example depicted how the changing healthcare delivery system is affecting providers thinking about their patients and processes. The fourth example addresses the consequences of success being achieved in the third example, namely, decreasing hospital revenues. This final example involves modeling and visualizing a large portion of New York City's healthcare ecosystem. Taken together, these four examples clearly illustrate how fundamental change can be addressed, hopefully successfully but sometimes with significant negative consequences.

OBSERVATIONS ON PROBLEM SOLVING

Table 10.1 summarizes the 16 case studies discussed earlier in terms of problems addressed and observations on the problem solving of the teams involved. The observations can be clustered into starting assumptions, framing problems, and implementing solutions.

Starting Assumptions

Who are the stakeholders in the problem of interest and its solution? It is essential that one identify key stakeholders and their interests. All critical stakeholders need to be aligned in the sense that impacts on their interests are understood. The automobile engine case study illustrates the consequences of not understanding these impacts.

Look at problems and solutions from the perspectives of stakeholders. How are they likely to be thinking? In the licensing technologies case study, it was crucial to understand buyers' exercise costs. This led to the idea that the licensor could provide the licensee consulting services to exercise the option less expensively.

Articulate and validate assumptions. Significant risks can result when there are unrecognized assumptions. The digital signal processor case study illustrated the importance of validating assumptions before deciding to invest in development. This can sometimes be difficult when key stakeholders "know" what is best.

Understand how other stakeholders may act. The effectiveness of a strategy is strongly affected by competitors. This is well illustrated by the case studies of the medical imaging system and the Affordable Care Act. Having one or more team members play competitors' roles can often facilitate this.

Framing Problems

Define value carefully. Translating invention to innovation requires clear value propositions, as illustrated by the R&D services and microprocessor case studies. In both cases, value needed to be framed from the perspective of the marketplace, not the inventors. Markets do not see their main role as providing money to keep inventors happy.

Think is terms of both the current business and possible future businesses. Current success provides options for future success, but perhaps with different configurations for different markets. The optical computing case study illustrated how current products and customers provide options for new products and customers.

TABLE 10.1   Observations on Problem Solving

Case Study	Problem Addressed	Observation
Products versus services	Better profit margins on services than products	Difficulty of changing business model
Defense conversion	Declining revenues from incumbent markets	Difficulty of changing business model
R&D as a service	Understanding what customers really value	Invention to innovation requires clear value
Academic strategies	Business plans that all constituencies support	Difficulty of keeping everybody supportive
Automobile engine	Reassessing merits of past investment decision	Need to align all critical stakeholders
Microprocessors	Understanding what customers really value	Invention to innovation requires clear value
Digital signal processor	Determining credibility of a priori assumptions	Validate assumptions before investing
Medical imaging system	Understanding likely strategies of competitors	Best strategy strongly affected by competitors
MRAM	Exploiting technology over many generations	Data integration increases confidence
Optical computing	Finding new broader market for technologies	Current success provides options for future success
Unmanned air vehicles	Replacing legacy solution with new technologies	Need to get legacies "off the books"
Licensing technology	Selling versus buying technology options	Crucial to understand buyers' exercise costs
Value opportunities	Exploiting market and technology opportunities	High payoff (and high risk) for innovation
Value deficiencies	Remediating poor performance and results	High costs/consequences for changing late
Population health	Making money keeping people healthy	Stratification and tailoring of processes crucial
Affordable care act	Enlarging market to leverage capacities	Best strategy strongly affected by competitors

254

Consider possibilities for customizing solutions for different customers and constituencies. The population health case study required stratification and tailoring of processes to varying health needs. This was critical to the viability of these population health offerings. Henry Ford, almost 100 years ago, was the last person to believe that everyone wanted exactly the same automobile.

Access and integrate available data sets on customers, competitors, technologies, and so on. The MRAM case study showed how data integration increases confidence. Great insights can be gained by mining available data sets, including internal sets, publicly available sets, and purchasable sets.

Plans should include strategies for dealing with legacies. The status quo can be an enormous constraint because it is known, paid for, and in place. The unmanned air vehicles case study illustrated the need to get legacies "off the books." Discarding or liquidating assets for which one paid dearly can be painful.

Implementing Solutions

It can be great fun to pursue market and/or technology opportunities. Innovators can earn high payoffs, albeit with high risks, as depicted in the value opportunities case study. The key is to have the human and financial resources to support and sustain the commitment to innovate.

In stark comparison, crises are not fun. The value deficiencies case study illustrated the high costs and substantial consequences of delaying change. Often, the status quo has devoured most available human and financial resources. When change is underresourced, failure is quite common.

The existing enterprise can hold change back. The products versus services and defense conversion case studies portrayed the difficulty of changing business models. New business opportunities may be very attractive, but if success requires substantially new business models, one should assess the enterprise's abilities to make the required changes.

Change should involve stopping as well as starting things. Stopping things will likely disappoint one or more stakeholders. The academic strategies case study illustrated the difficulty of keeping everybody supportive. The consequence is that the status quo dominates, especially when senior management team members were recruited to be stewards of the status quo.

RESEARCH ISSUES

This book has presented a methodology for addressing complex systems and enterprises. We have provided linkages to the historical roots and technical underpinnings of this methodology and outlined a catalog of component phenomena for populating visual and computational representations of complex systems. Thus, the methodology rests on an impressive body of knowledge.

However, there are several fundamental issues that need to be addressed for this endeavor to mature and be widely employed by both researchers and practitioners. The issues and questions outlined in this section need to be resolved if visualization and computational modeling are to move beyond the current state of each instantiation of this approach being an idiosyncratic and often heuristic creation by modelers who are often unaware of the foundation upon which they can build.

Decomposition

The starting point for visualizing and modeling is the decomposition of an overall phenomenon, for example, healthcare delivery, into component phenomena at varying levels of abstraction and aggregation. A central question at this point is what phenomena belong in each of the multiple levels? Further, how does the representation of each phenomenon depend on its level? This can be addressed in part by considering natural part–whole relationships. However, the question remains what wholes and what parts are needed to address the question that motivated the modeling initiative?

As an aside, it is important to reemphasize the need to be very clear at the outset about what questions the model is intended to address. When the only objective is to develop a model, and the questions to be addressed remain elusive and ambiguous, the natural tendency of modelers is to include everything they can imagine that will ever be needed. The result tends to be an unwieldy composition that is difficult to understand and maintain.

Mapping

The choice of which phenomena will be represented at each level leads to choices of how to characterize each phenomenon. The catalogs of models discussed throughout this book can be quite helpful in this regard. Each representation will have defined input–output variables. The next question is what variables cross the levels between representations? Further, what transformations are needed to connect across levels? Basic issues here include units of measure, coordinate systems, and time. Zeigler et al. (2000) addresses these issues within the context of dynamic systems.

It is important to note that resolution of these basic issues is necessary but not sufficient for assuring Tolk's (2003) semantic interoperability. Being able to connect two models and have them jointly compute some outputs does not assure that these outputs are valid and meaningful. The issue here is one of "assumption management." Are the assumptions of the two or more interconnected models compatible? This is straightforward when the modelers are the creators of all the component models, but far from easy when some of the component models are legacy software codes.

Scaling

It is often the case that models begin with only a small number of simulated agents (e.g., patients) or only a fraction of the entire transportation network that is of interest. The intention is to scale up such smaller models to address the whole problem of interest once experience and confidence is gained with the initial models. Scaling is often very difficult and results in large unfathomable models that compute very slowly. The modelers can lose any intuitions of what is happening in the scaling process.

The first question is, given the targeted scale of the modeling effort, what should be the unit of scale for each phenomenon? A related question is by what quantum does each unit scale? Perhaps millions of patients are better simulated as cohorts rather than individuals. Perhaps the flow of thousands of vehicles should not start with the dynamics of each vehicle, but instead consider waves of vehicles.

Approximation

The creation of large multilevel models inevitably requires using approximations. The central question is what approximations are best used for data and computational efficiencies? For example, what probability distributions are used for arrival and service times in a discrete event simulation? Triangular distributions might be much easier to implement than lognormal distributions, compute much faster once implemented, and, very importantly, have parameters that are much easier to estimate.

One needs to ask about the implications of different choices? For example, defining agents as cohorts of patients rather than individual patients will reduce the variability across patients. If this variability is the primary issue of importance, some countermeasure for the variance reduction may be needed. In general, small-scale examples can often be used to gain understanding of how the effects of approximations propagate.

Identification

The question of interest here is how can structural properties of processes be inferred from design and operational data sets? This is important because many complex systems have no "as is" blueprints, that is, such systems emerged rather than being designed. Instead, one may have data on millions of transactions throughout the system's processes. One needs to have algorithms that can infer processes from such data sets, often without any baseline process maps to help with validation.

This raises the question of what are the best metrics for characterizing the "fit" of an inferred process to a data set. For identifying input–output relationships, one could use mean-squared-error as the metric. However, for process maps where relationships among nodes are the concern, one is fitting networks to data rather than equations. In this case, one might use something like percentage of empirical relationships captured by the network representation.

Parameterization

Structural representations of processes will usually have parameters such as rates, means, and probabilities that need to be estimated. In order to estimate these parameters, one first has to address the question of how data sets can be accessed and normalized across elements of the enterprise. For example, within healthcare, how can clinical, financial, and claims data sets be combined, while maintaining patients' identities and rationalizing varying time scales?

Once data sets are combined and rationalized, how can unbiased parameter estimates best be obtained from the integrated data set? A key issue is assuring that the data set used for estimating parameters is representative of the population for which predictions are sought. A more refined concern is estimating parameters for baseline "as is" systems versus potential "to be" systems.

Propagation

Structural and parametric uncertainties can have far-reaching effects as they propagate across representations and levels of the overall model. This raises the question of how uncertainties can best be propagated across multiple representations at multiple levels. In particular, how is the variability associated with one level propagated to other levels when simple propagation of point estimates is unwarranted?

Various mechanisms might be adopted, but how are levels of variability attenuated or accentuated by different approaches to propagation? The key issue is that approximations can have effects beyond the immediate impacts motivating the approximations. This can be a rather complicated issue and have significant higher-order effects and unintended consequences.

Visualization

If models are to be used to support a wide range of decision makers, the model outputs have to be accessible by people who are far from modeling and simulation experts. This raises the question of how the "state" of a multilevel system can best be characterized and portrayed. The answer to this question should be determined by the nature of visualizations most meaningful to the key stakeholders in regard to the questions targeted via the multilevel model.

Beyond portraying the state of the system, stakeholders are often concerned with the nature of relationships between levels of the model. How can the relationships within a multilevel system best be portrayed to enable experimentation and insights? This question concerns how best to enable stakeholders to manipulate the relationships between levels of the overall model. Once stakeholders are "in the loop" of choosing assumptions and manipulating parameters, stakeholder buy-in is usually greatly enhanced.

Curation

This book, as well as Rouse and Bodner (2013), outlines a wealth of component models for potential inclusion in multilevel models. However, this wealth is not often used. This is due to both a lack of knowledge of these resources and difficulty in accessing them (Rouse & Boff, 1998). Professionals in modeling and simulation seldom access the academic journals originally reporting these models. Even when practitioners are aware of such publications, they are seeking computer codes, not research treatises.

How can component models be represented, archived, maintained, and accessed to facilitate rapid model integration? Put simply, these resources need to be curated. There needs to be one point of access for the many hundreds of models discussed in this book and by Rouse and Bodner. This access should enable downloading computer codes, documentation on assumptions and use, and original reports of the development and validation of these models. Of course, this begs the basic question of how participating organizations can be incentivized to contribute to and make use of the curated archive.

CONCLUSIONS

Addressing complex systems such as healthcare delivery, sustainable energy, financial systems, urban infrastructures, and national security requires knowledge and skills from many disciplines, including systems science and engineering, behavioral and social science, policy and political science, economics and finance, and so on. These disciplines have a wide variety of views of the essential phenomena underlying such complex systems. Great difficulties are frequently encountered when interdisciplinary teams attempt to bridge and integrate these often-disparate views.

This book is intended to be a valuable guide to all the disciplines involved in such endeavors. The central construct in this guide is the notion of phenomena, particularly the essential phenomena that different disciplines address in complex systems. Phenomena are observed or observable events or chains of events. Examples include the weather, climate change, traffic congestion, aggressive behaviors, and cultural compliance. A team asked to propose policies to address the problem of overly aggressive motorist behaviors during inclement weather in the evening rush hour might have to consider the full range of these phenomena.

Traditionally, such problems would be decomposed into their constituent phenomena, appropriate disciplines would each be assigned one piece of the puzzle, and each disciplinary team would return from their deliberations with insights into their assigned phenomena and possibly elements of solutions. This reductionist approach often leads to inferior solutions compared to what might be achieved with a more holistic approach that explicitly addresses the interactions among phenomena and central trade-offs underlying truly creative solutions. This book is intended to enable such holistic problem solving.

We conclude with the themes with which this book began:

- Understanding the essential phenomena underlying the overall behaviors of complex systems and enterprises can enable improving these systems.
- These phenomena range from physical, behavioral, and organizational, to economic, political, and social, all of which involve significant human components.
- Specific phenomena of interest and how they are represented depend on the questions of interest and the relevant domains or contexts.
- Models from systems science and engineering, defined *very* broadly across disciplines, can provide the necessary understanding.
- Manifestations of such models can range from static visualizations to interactive visualizations to deep computational realizations.

These themes collectively can enable supporting creative and successful solutions of problems associated with complex systems and enterprises.

REFERENCES

Rouse, W.B. (1992). *Strategies for Innovation: Creating Successful Products, Systems and Organizations*. New York: Wiley.

Rouse, W.B. (1996). *Start Where You Are: Matching Your Strategy to Your Marketplace*. San Francisco: Jossey-Bass.

Rouse, W.B. (1998). Computer support of collaborative planning. *Journal of the American Society for Information Science*, 49 (9), 832–839.

Rouse, W.B. (Ed.). (2006). *Enterprise Transformation: Understanding and Enabling Fundamental Change*. New York: Wiley.

Rouse, W.B. (2011). Necessary competencies for transforming an enterprise. *Journal of Enterprise Transformation*, 1 (1), 71–92.

Rouse, W.B. (2014). Human interaction with policy flight simulators. *Journal of Applied Ergonomics*, 45 (1), 72–77.

Rouse, W.B., & Bodner, D.A. (2013). *Multi-Level Modeling of Complex Socio-Technical Systems (RT-44 Phase 1 Report)*. Hoboken, NJ: Stevens Institute of Technology, Systems Engineering Research Center.

Rouse, W.B., & Boff, K.R. (1998). Packaging human factors for designers. *Ergonomics in Design*, 6 (1), 11–17.

Rouse, W.B., & Boff, K.R. (2004). Value-centered R&D organizations: Ten principles for characterizing, assessing, and managing value. *Journal of Systems Engineering*, 7 (2), 167–185.

Rouse, W.B., & Howard, C.W. (1995). Supporting market-driven change, in D. Burnstein, Ed., *The Digital MBA* (pp. 159–184). New York: Osborne McGraw-Hill.

Rouse, W.B., & Serban, N. (2014). *Understanding and Managing the Complexity of Healthcare*. Cambridge, MA: MIT Press.

Tolk, A. (2003). The levels of conceptual interoperability model. *Proceedings of the Fall Simulation Interoperability Workshop*, Orlando, Florida, September.

Zeigler, B.P., Praehofer, H., & Kim, T.G. (2000). *Theory of Modeling and Simulation: Integrating Discrete Event and Continuous Complex Dynamic Systems*. New York: Academic Press.

INDEX

Modeling and Visualization of Complex Systems and Enterprises:
Explorations of Physical, Human, Economic, and Social Phenomena, First Edition. William B. Rouse.
© 2015 John Wiley & Sons, Inc. Published 2015 by John Wiley & Sons, Inc.